BLACK HEREFORDS

BLACK HEREFORDS

J&N

The Birthplace

Norma Decker Hoagland

PELICAN PUBLISHING
NEW ORLEANS

The word "Pelican" and the depiction of a pelican are trademarks of Arcadia Publishing Company Inc. and are registered in the U.S. Patent and Trademark Office.

ISBN 9781455628599

Printed in the United States of America

Published by Pelican Publishing
New Orleans, LA
www.pelicanpub.com

To our children, Anne, Dirck, and Emma,
who had to put up with debates about business and politics
every night at the dinner

CONTENTS

FOREWORD

New major cattle breeds don't come around very often. Most are combinations of existing breeds, but rarely do they achieve international recognition as a uniquely separate cattle breed. There have only been four major beef cattle breeds that started in America. No, the Texas Longhorn is not one of them. Longhorns were brought to the New World by the Spanish Conquistadores in 1512.

Hereford cattle originated in Herefordshire, England, in the mid-1700s. They were brought to America in 1817 by the American politician Henry Clay of Kentucky. A genetic mutation produced a hornless variant in Herefords, which was used to create the first separate American breed in 1889. Iowa cattle rancher Warren Gammon started the registry in 1910 known as the American Polled Hereford Association (Polled meaning without horns).

The next cattle breed developed in America was started by Robert Kleberg of the King Ranch in 1918. Their new Santa Gertrudis breed combined Shorthorn and Brahman cattle. It wasn't until 1940 that the King Ranch was officially recognized as producing a new cattle breed by the United States Department of Agriculture.

In 1912, the US Department of Agriculture suggested a three-eighths Brahman and five-eighths Angus cross as a hardy breed that would be particularly resistant to high temperatures. In the 1930s, the cross was used by several independent individuals. The International Brangus Breeders Association was formed in 1949, making Brangus the third American beef breed.

Fourth would be the Lasater Ranch in South Texas, developing the Beefmaster as a new cattle breed in 1931. Tom Lasater crossed Milking Shorthorn, Hereford, Brahman genetics to create an animal called the Beefmaster. It wasn't until 1954 that it was recognized as a specific breed of cattle.

The fifth American cattle breed was the Black Hereford created at J&N Ranch in Leavenworth, Kansas, by Joe and Norma Hoagland in 1997. It was recognized by the Food and Agriculture Organization of the United Nations, based in Rome, Italy, in 2003.

This is the story of how my wife and I created the Black Hereford cattle breed. Today it is the fastest growing cattle breed in America with over six hundred members in forty states and several other countries.

As the name J&N suggests, it took a partnership. Being married to one of the most beautiful women on the planet was helpful. It turns out she was also very intelligent. She quickly understood people and complex issues. At five foot nine with long black hair and green eyes, when she walked into a room, she commanded attention. As the trailing spouse, I quickly learned to take advantage of how we were perceived as a couple and what we could accomplish together.

As a biographer and author, having already published two history books, Norma tells our story.

Joe Hoagland

BLACK HEREFORDS

CHAPTER ONE

FAMILY LESSONS FOR LIFE

The first Black Hereford calf was born on our J&N Ranch, which runs along the Missouri River, just north of Kansas City, on the Kansas side. Joe and I settled on that name for our ranch in 1978 (and my mother always said it should be N&J Ranch!), but our family's farming interests go back a very long way. Early on, we realized we shared a love of land, the heart of a farmer.

We set up J&N Ranch, LLC as a corporation to farm and ranch and to own and transfer land from one generation to the next. We had seen smooth and rocky inheritance of ag land and wanted ours to be easy. Joe always looked ahead and had the skills for us to navigate this. Land values continued to increase throughout our working lives. You know the old saying, that "farmers live poor and die rich." As a family, we have tried to avoid ruinous estate taxes by incorporating our life's work in agriculture. Some of our land has been in the family since 1866—a long time by American standards. We wanted to be sure it could continue, if possible.

People are all the sum of our life experiences. What we learn from those times, both good and bad, gives us tools to go forward. This is a story of Joe's and my experiences in the cattle industry, but also in politics, advertising, law, beauty pageants, baseball, modeling, and farming. All of these threads and stories may seem separate, but woven together, they gave us the strength and resources to start a new cattle breed. It took years to get this goal accomplished, and there were times, looking back, when it nearly didn't happen. Believe me, all our skills and hard work were needed to accomplish our goal of starting the Black Hereford breed and member-owned association.

Joe's family, the Hoaglands, and my family, the Deckers, are both Dutch and among America's oldest. We are, in fact, distant cousins, both direct descendants of the first white child born in New York, Sarah Rapalje, whose parents arrived in New York in 1624.

The Deckers were Protestants at a time in Europe when it was predominately Catholic, and they were in danger. Battles over religion are nothing new, but the promise of religious freedom was enough for my ancestors to pack up and leave everything familiar and become settlers in the New World.

Dirck Jansen Hoogland, Joe's first ancestor who arrived in the New World in 1657, was not a penniless immigrant. His father owned the ship in which he traveled across the Atlantic. In the Netherlands, in the town of Leyden, there is a Hooglandse Kerk (church), Hooglandt Street, and Hoagland café. Dirck left Holland to seek trade in beaver pelts to ship back to Amsterdam to make hats. It was a financial success. He built the first stone building on the island of Manhattan and owned forty acres of land north of today's 44th Street running along the East River.

These ancestors awe us with their courage and hard work.

Still speaking Dutch for several generations after British rule began in New York in 1675, the family names were anglicized in official records. Hoagland was spelled first as Hoogland, then Hogeland, Hoghland, Hoagland, and Hoghlant. Decker, my maiden name, was spelled Dekker, DeDekker, or DeDecker often in those days.

Our families farmed or were merchants in the New York and New Jersey area for decades. Several of our ancestors fought in America's Revolutionary War against the English. They were patriots fighting for the freedom their ancestors had come to America for. Alexander Hamilton was buried next to Dirck's grandson, Adrian Hoghland, in the graveyard of Trinity Church in New York.

To this day, the old Dutch families in New York still know each other. Names like Decker and Hoagland are part of The Holland Society of New York, established in 1885. It is a historical and genealogical society only for those descended from the original settlers of New Amsterdam (today's New York). Joe's father was a member, and Joe served on their board of directors for many years.

Our family's contribution to America has been celebrated by the Dutch. In 2009, to celebrate the four-hundredth anniversary of Henry Hudson's voyage of discovery to the New World and the contribution of these historic Dutch families in America, Joe's picture was displayed in the Amsterdam city hall and featured on the cover of the Dutch magazine *fd* to celebrate the Dutch influence on America.

From these brave pioneers, both of our families came to Kansas as anti-slavery abolitionists during the Civil War period in America. After the Kansas-Nebraska Act of 1854, the Dutch communities in the Northeast encouraged people to come to Kansas, to help vote it into the Union as a Free State. Levi Decker, my great-grandfather and George Hoagland, Joe's great-great-grandfather, both served in the Union army during the Civil War. Levi served in the twenty-seventh New Jersey Volunteer Infantry. George is listed as a blacksmith

Adriaen Hoghland, a grandson of Dirck Jansen Hoagland, one of our ancestors in New York. Alexander Hamilton is buried next to him in the cemetery at Trinity Church, 89 Broadway, New York, NY. His gravestone reads: "Here lies the body of Adriaen Hoghland who departed this life the 28th of July 1772 Aged 73 years."

on his discharge papers from the Grand Army of the Republic. Both stayed in "Bleeding Kansas" after their military service and lived only a few miles from each other.

Once in Kansas, Levi homesteaded a farm in Tecumseh, just southeast of Topeka, the "free state" capital. George stayed in Kansas, also, and is listed as a minister for the Evangelical United Brethren Church in Topeka. That building was demolished to build Highway I-70 through downtown Topeka. A coincidence is that this was the church my family attended when I was a child. We wonder if our families knew each other way back then and think they likely did.

Both Joe and I had grown up around agriculture; I on Levi's dairy farm in

Joe on the cover of the Dutch magazine fd on December 12, 2009, for an article featuring Americans who are descendants of the original Dutch colony of New York.

Tecumseh, and Joe on properties his father owned in Kansas and Missouri.

After service in WWII, Joe's parents bought land in what was then a rural area of Overland Park, a small town south of Kansas City. Although they had neighbors, Joe had lots of land to roam on and lots of animals to grow up around and care for: horses, cattle, donkeys, ducks, and dogs.

Joe's mother, Ruth Reed Hoagland, understood her very active son and gave him lots of freedom. She even cleared a place and built a small baseball field for Joe and his friends to play ball on, along with Joe's sister Ruth Ann. A posse of boys, Joe, Dave, Mike, and Lee, would pack a PBJ sandwich and head out on their bikes for adventure. This was a gang of boys who explored on their bikes clear out south to the nearby town of Olathe—a distance of about fifteen miles! That freedom, given by his mother, gave Joe opportunities to explore, dream, and think for himself and develop confidence. She is largely responsible for his confidence and for his artistic side, in my opinion, as she was a successful self-taught artist.

From his father, Virgil Hoagland, Joe got different gifts. Virgil loved to travel, and Joe had experiences from different places at a young age, Canada and Mexico, as well as trips around the US. Nearly all of these were business trips because Virgil liked to take his family along, something Joe later did. Also from his father, Joe learned hard and dedicated work was how you showed your love

Joe on his pet donkey, growing up on his farm in Overland Park, Kansas.

for your family. No, I'm not saying that just right, having known Virgil. Hard work was not only how you showed your love for your family, but your respect for yourself.

Virgil was the fourth generation in his family to run their family business, Racket Merchandise Company. It was a hardware and housewares supplier with several locations in the Kansas City area. Racket has an interesting history of its own, but we mustn't get ahead of our story. The point here is that Joe saw his father work tirelessly, seven days a week. However, in the family we joked that he only worked from 10:00 a.m. to 2:00 p.m. on Sundays! Holidays were even busier at the stores. Virgil had a hard time talking about anything other than work, which I learned when I joined the family. Work was the center of their family's life as Joe grew into manhood. He began to work in the business as a young boy, doing simple jobs in the warehouse to help.

Work was a critical connection between Joe's childhood and my own, a value at the core of our individual selves and later, of our marriage. We worked, and still work, as a team.

Growing up on a dairy farm, as you can imagine, work came first in my family, too. The cows were fed before the people were fed. Chores had to be done, and work was the glue that held everything together. I saw my parents working together to sustain our family of five on our small farm. My parents taught by example and little did I realize how valuable that lesson of marital partnership would be in my life.

Our Decker Farm is where my great-grandfather Levi Martin Decker settled in the early post-Civil War days of 1866. From Levi, the farm came to my grandfather, Calvin Decker and his wife Pearl Peck Decker. My father was born in the house they built at the time of their marriage in 1911, the same farmhouse I grew up in. (Still with a functioning outhouse!)

Tecumseh was our nearest town, just east of Topeka, the capital of Kansas. Tecumseh is also near Lecompton, which was the early "slave" capital, established by the pro-slavery forces during the Civil War. Topeka and Lecompton were only twenty miles apart, with Tecumseh nearly in the middle of these warring factions. In 1861, Kansas was voted into the Union as a free state and Topeka became the permanent capital. So truly our land was in the area known as "Bleeding Kansas," positioned right between the pro-slavery and anti-slavery capitals before statehood.

We take pride in the fact that both of our families, mine as well as Joe's, migrated to Kansas from the northeast of America to vote in Kansas as a free state. Again, our Dutch ancestors were abolitionists from those churches in New York and New Jersey that promoted moving west to settle Kansas as a state free from the evil of slavery.

Norma's great-grandfather Levi, front row, second from left, with his children. Norma's grandfather Calvin is in the back row, second from left.

From Levi Martin Decker to his son Calvin, to my father, Dayton, came the farmland in Tecumseh. Levi had given a portion of that land and built a school for the children in our area of Tecumseh in 1881. Education was always a priority in my family, passed down through the generations.

Decker Farm had a dairy herd, and we raised alfalfa, corn, and oats. On that land, I had the opportunity, much like Joe, to grow up in pastures, woods, and streams. While I had no neighbors for playmates, I had my imagination to keep me company. Memories of searching for a litter of kittens in the chicken house and making my own playhouse in our big haybarn are the earliest things I remember. My favorite sanctuary was my mother's rock garden surrounded by lilac bushes, so deliciously scented in the springtime. Lilacs are still my favorite flower.

Living on a dairy farm, we drank lots of milk and ice cream was a favorite treat. My mother's specialty became cream pies: coconut, chocolate, and butterscotch. We had a huge garden and grew nearly everything we ate, including home-grown beef. My father always had a few Hereford steers he raised for beef for the family, along with our Holstein and Brown Swiss milk cows. I helped with my share of chores and vividly remember putting up hay in the summer—small bales—lots of them!

Decker School, built in 1881 by Levi Decker. It had a well water pump in front, an outhouse, and shelter for horses at the back. It is now a private residence on 29th Street, one mile east of Tecumseh Road, near Topeka, Kansas.

Summers were busy with strawberries, peas, green beans, tomatoes, and potatoes. We froze and canned our produce, pickled beets, and cucumbers, working in the summer heat. There were grapevines for grape jelly and fruit trees for pears and peaches. I can still picture our root cellar, lined with those beautifully colored canning jars. The work was always worth it when we could open those treats in the cold of winter.

My mother, Jennie Mae Smith (Jeffery) married my father, Dayton Lewis Decker after her first husband George Wayne Jeffery died of polio shortly after his service in WWII. After their wedding, Mom moved to our Decker family farm in Tecumseh, bringing her two small children, Carol Jeanette and Larry Wayne. I was born a couple of years later and grew up as a much younger sibling.

I was a pesky youngster to my siblings, and I have proof. Often, I was sent out in the pasture, across the creek to "bring the cows in for milking." Now, the cows would come to the barn all by themselves, but I thought I was playing an important role on the farm. Only when I got a little older did I realize with a flushed face that this was a ploy to get rid of me for a while! But I played solo, sang and danced to the trees, and invented games for myself. I grew oversized confidence in myself and was largely unafraid of the future, trusting the world would take care of me as my family and farm had.

It was a sweet time to grow up, in the 1950s, for both Joe and me. People trusted each other, children were allowed to roam and play outdoors with no fears of strangers or dangers. On warm days, we both remember running outside to play, hearing our mothers say, "Come back in time for lunch!" I believe this freedom gave us the mindset that we could accomplish anything we set our minds and hearts to do.

Taking a break from farm work, Norma's father, Dayton, takes her brother Larry and little Norma for a ride.

Growing up in the youth organization of 4-H added greatly to the experiences of my early years. My parents started a club for our area the year I was born, Decker 4-H Club, which is still in existence today. In 4-H, we learned about different project areas, kept records of our learning and progress, and then learned to speak to our club members about our projects. I learned also that I loved to speak to a group. The Shawnee County Fair, and if our projects were judged good enough, even the Kansas State Fair, were exciting times to show our projects, meet new people and compete.

Shawnee Heights High School for me was a wonderful time. I participated in theater, music, forensics, debate, cheerleading, and more. It was a small school, so I was able to try any activity I wished. My dear friend, Ann, once told me that I "had more confidence than sense." Ready to take on the world, I applied for scholarships to the University of Kansas.

Joe went to school in the rapidly growing Kansas City suburb of Overland Park, Kansas. Although he lived on a large acreage, he saw the nearby housing boom of post-World War II firsthand. Both of our fathers were veterans, army and navy. Joe remembers 75th Street in Overland Park being a gravel road, something so difficult to think of in 2026! Joe's education began at Antioch Elementary School but with difficulty not fully understood in 1952. Joe has dyslexia, a learning disorder that made learning to read very difficult. He was held back in school, taking the second grade twice. This became his lifelong challenge to overcome—and he did. Again, hard work and confidence were his allies.

After high school, Joe traveled back east on a football scholarship. His belongings were packed into a steamer trunk and shipped off. Due to an airline strike, he took the train to New York. After a year there with his many distant cousins, he transferred to Tulane University in New Orleans. His three years there were marked by witnessing the first true racial discrimination he'd seen. Raised as a Northerner, he was taken aback by separate entrances at restaurants, bars where African Americans could only order drinks through a window, and Black passengers moved to the back of busses. Tulane was a rigorous academic school, but he made good grades and became the president of his fraternity. After college, he was ready for any challenge facing him. Majoring in history, law school was already on Joe's mind.

My college experience was very different. Attending the University of Kansas (KU) on scholarships, my first two years were marked by political and social unrest. My freshman year had Vietnam peace marches on campus and civil rights protests in the city of Lawrence, Kansas. The night of the bombing and fire in our KU Student Union building is still a vivid memory for me.

From my all-women's scholarship hall, Watkins Hall, less than a block away from the Student Union, we could see and hear it all. We girls took sandwiches and coffee to the firefighters and students that helped save the building. When a curfew was set on campus because of the unrest, I had the eerie experience of having a very young National Guardsman, with a rifle, stationed just outside my room's window. I wondered if he was more frightened than I was. Students were sent home near the end of that semester without taking our final exams. This wasn't a good way to begin a thoughtful learning experience! After two years at KU and living at Watkins Hall, I decided to work and get away from the turmoil. I needed time to decide what major I wanted to pursue at KU.

There was also another kind of personal turmoil that I was fleeing from at that point in my life. While living at Watkins Scholarship Hall, I had signed up for a contest at KU called the Best Dressed Coed. I was quite confident in my ability to sew my own clothes, being active in 4-H, where I would sew and then model my finished garments for a judge. I didn't view this as a beauty pageant, but a test of skill. I won that title of Best Dressed Coed for myself and my scholarship hall in 1969, my first year at KU.

After that title, my roommates, Ann and Debbie, signed me up for a beauty pageant! I returned to the hall one evening and they proudly announced this to me. Being a contestant in a beauty pageant had never been a goal of mine, but it sounded like fun, I naively thought. So, I packed a small bag and went to Kansas City to enter the Miss Kansas USA pageant of 1970.

From my years of 4-H in clothing construction, I had sewn clothing and then modeled my garments for judging. So, walking on a runway was no problem for me, although I had never done it in a swimsuit! The personal interviews were interesting and fun. The final portion of the competition was impromptu or extemporaneous speech. Having taken debate in high school, I was able to quickly formulate a short speech. Remember, these were the Vietnam War years, so I spoke about the difficulty between the generations in their perceptions of that war. I was likely not the most beautiful candidate, but I believe I won on the strength of my speaking ability.

My experiences following this perceived victory were not so pleasant, however. The couple who owned the franchise for the USA Pageant were not happy

Norma, newly crowned Miss Kansas USA 1970 and the runners-up that year.

with me and did little to help prepare me for the Miss USA Pageant that followed. My physical appearance, five foot nine inches and thin, was criticized at length. They fitted me with special enhancing bras. I was instructed to diet and take a huge bottle of tablets labeled, "Extract of beef mammary gland." I did not take them. At the time, I blamed myself for not living up to their expectations, as a young inexperienced girl would. Interestingly, forty plus years later I was confronted by a woman at a wedding reception in Kansas City who approached me and asked if I was Norma Decker. When I replied, "Yes," she then angrily announced she "was supposed to win" the Miss Kansas title that year, because her parents were friends of Bob and Jacquie, the pageant owners! The mystery of why I wasn't liked or prepared was finally solved, but too late for my unhappy year.

So many years later, the Miss USA Pageant of 1970, held in Miami, is still an unpleasant blur. Most of the contestants were young women whose life goal was to be in that pageant, and they intended to do everything possible to win. My roommate was a Georgia peach whose lush accent stopped when we entered our hotel room alone. No one was friendly, helpful, gave advice or encouragement, with one exception. I was taught that a long strip of masking tape, applied carefully, could give an illusion of cleavage where there was little. This was a lesson I never used again.

Beauty pageants were not a part of the friendly world I was naively accustomed to, back in Kansas. A life lesson of competition, hypocrisy, and meanness was learned. I barely survived and returned home with my spirit dragging behind me. Lessons about people learned.

When I returned to campus for my year as Miss Kansas, there were pageant publicity events I was to attend. Most of these, like cutting ribbons to open a shopping center, were fun but made attending college classes more difficult. However, it was the constant unwanted pressure from men that began to be oppressive. Even on campus, pressure for dates was difficult to cope with. This, along with the campus unrest I previously described, made one thing clear to me. I needed to escape. A graduate student I knew at KU provided the answer.

Helping Lois Dixon with data collection needed for her PhD, I moved to and worked at the Parsons State Hospital for Retarded Children. Working with those children was both therapeutic and satisfying for me. Lois successfully defended her thesis, and I decided it was time to get back to my education. I had lost my scholarships by leaving KU for two years, but by working at various jobs, I was able to support myself and work toward my degree in Human Development and Family Life. I planned to be a licensed preschool teacher, able to develop school curriculum and plan nutritious meals. I moved back to Lawrence, Kansas, back to KU.

Group photo from the Topeka Capitol-Journal, May 14, 1970. Norma is pictured in the center along with contestants from Louisiana, Arkansas, Missouri, and Texas at the Miss USA Pageant in 1970.

Joe had returned to Overland Park from Tulane University, diploma in hand, ready to help his father run the family business, Racket Merchandise Company. When he was once again put to work in the warehouse, just as he had done as a young boy, he and his father disagreed about his role in the family business. Joe decided to get a different job, away from the family, and applied to law school. Virgil replied that all was fine with him, but he was not going to pay for law school. The gauntlet was thrown down.

Coming from a Republican background, Joe found a staff job working to help elect Kent Frizzell who was running for Governor of Kansas in 1970. Working with other bright young men in those political circles made an impression on Joe. He met US Senator Bob Dole. It was the 1970s, and as a young, idealistic, and motivated person, he began to think of what he could do with his life. Another pressing issue was: how to pay for law school?

Joe had applied, was accepted, and began at KU's School of Law in the summer of 1971, with politics in the back of his mind. In Professor Barkley Clark's class on statutory construction, or how to write laws, Joe did an internship for Kansas State Senator Norman Gaar of Mission Hills. A fellow KU law student, Richard Walker of Newton, Kansas, had a similar internship with Senator Joe Harder. Richard and Joe became friends, sharing experiences at the Topeka State Capitol during their internships.

Richard decided to run for public office against his local state representative in 1972, inspiring Joe to do the same. Getting campaign advice from another young state legislator at the time, Paul Hess of Wichita, both Joe and Richard won their respective legislative seats. With pay of ten dollars per day and twenty-five dollars per day for expenses, law school became affordable! Joe flourished with the work of policy decisions and gained the respect of his constituents who had elected him from his hometown of Overland Park.

This was a different type of political environment than we think of today in 2026. The vast majority of senators and representatives were World War II veterans, much older than Joe and Richard. The tone was civil. You could disagree on an issue and still be friends. Cooperation between parties was the norm. Issues tended to be urban vs. rural, not based on party affiliation. Votes were based on what was good for the people you were representing. It was a responsibility that Joe, even as young as he was, took very seriously. He excelled and although he was still in law school, he was assigned to the Judiciary Committee, usually reserved for practicing attorneys. Joe would soon move into the leadership in the House of Representatives.

Politics became his life. Researching topics, working with others who perhaps didn't agree with him, making decisions with his constituents in mind, all became his work and came naturally to him. This political life extended well

JOE HOAGLAND IS QUALIFIED!

EXPERIENCE

- Resident 24 years.
- Successful businessman.
- Graduate Shawnee Mission West High School.
- Graduate Tulane University.
- Past Vice-Chairman of ALSAC Leukemia Crusade.
- Law Student.
- Past Bill Drafter and Research Clerk for the Local Government and the Federal and State. Affairs Committees of the Kansas Senate.

Joe's first campaign for the Kansas State Legislature in 1972 resulted in victory—the first of six consecutive winning elections.

into the next phases of Joe's adult life. Lessons learned here, in politics, added to his toolbox of experiences.

I am still amazed that Joe, with his dyslexia, could manage law school with the extensive reading required and pass the Kansas Bar Exam in 1974. All of this was done while serving as the State Representative of District 22 of Overland Park, commuting from there to law school at KU, in Lawrence, and to the Capitol in Topeka.

In February of 1974, I had been living back in Lawrence for a while, taking classes again at KU and working to support myself. I got a call from Nancy Archer, a friend I had lived with at Watkins Scholarship Hall. She wondered if I'd like to go on a double date with my friend and fellow "Watkins Girl," Susan Harshaw, who was dating Richard Walker. My date would be Richard's friend, a law student who was already serving in the Kansas House of Representatives. Although I'd sworn off dating for a long time, I thought a social event of the Kansas Legislature would be an interesting evening. It was Joe. He proposed on our third date. We were married in August of the same year. When you know it, you know it.

CHAPTER TWO

POLITICS AND LAW

Before I even met Joe, he had already made a name for himself as an advocate for consumer protection. Looking out for the rights of everyday folks became a theme for him. In his very first session in the Kansas State Legislature, he introduced a bill to create a small claims court where people could go to settle disputes involving $300 or less without having to hire a lawyer.

To explain his actions, Joe was asked to write an article for the *Kansas Bar Journal.* This appeared in the Bar Journal of the Kansas Bar Association, Spring of 1973, pages 13-17. It was the first time the Bar Journal Board of Editors allowed a law student to be published.

Our friend Richard Walker had left the legislature after three terms. He went on to become a district judge and eventually retire from the Kansas Court of Appeals before becoming president of the Kansas Historical Society. It was in that latter capacity that Richard interviewed Joe for the Kansas Oral History Project in 2023, and he spoke about the small claims court that benefited everyday Kansans.

As Joe describes it in that interview:

"[Small claims court] was designed to give a remedy to people that had a civil complaint oftentimes involving faulty construction work or some relatively small transaction that there was no remedy for. You couldn't afford to hire a lawyer. The lawyer fees were more than what you were wanting to recover, and this gave them a way out."

People would fill out short, easy paperwork to get access to the courthouse, and have a judge help them get justice for a small claim. There were relaxed rules of evidence you had to provide. It proved to be a great success. From its inception until today, the limit for recovering an injustice has risen from $300 to $10,000.

Imagine, still being a law school student and having the confidence and

A Look at the Kansas Small Claims Procedure Act*

By JOSEPH J. HOAGLUND

The 1973 session of the Kansas Legislature took progressive steps in the area of consumer protection. Among the most significant was the enactment of Senate Bill 62, the small claims procedure act. As distinguished from a small claims court, Senate Bill 62 provides an alternative procedure for handling small claims (those involving $300 or less) in the courts of limited jurisdiction throughout the state.

THE NEED FOR A SMALL CLAIMS PROCEDURE

The rationale for passage of a new procedure to be incorporated into the present code of civil procedure is analogous to a person cutting his finger while peeling carrots. Here, the procedure toward restoration is simple. He applies a bandaide to his injury. If the procedure were so complex that only a surgeon could skillfully administer it, the injured party might not choose to resolve the harm. The relief to his finger is simply not worth the expense of employing an expert. Clearly, there is no need for the complex surgical skills of a practitioner for such a slight injury. Yet, a simple cut could become infected if proper care is not taken.

For a plaintiff in a small claim dispute, the procedure toward restoration is not simple. Our complex rules of civil procedure and evidence tend to necessitate the aid of a skilled expert. However, since the amount of legal fees often consume a large part of the amount recovered, it is not economically feasible for most persons to retain an attorney in a small claims case, nor is such litigation profitable for most attorneys. Therefore, these disputes go unresolved. When proper care is not given to a cut finger, the result is the risk of infection. For a plaintiff with an unresolved small claim, the result is a deep sense of frustration with our system of justice and equity. Like an injured finger, a small claim

* Effective July 1, 1973

Kansas Bar Journal, *Spring 1973, page 13. "A Look at the Kansas Small Claims Procedure Act" by Joe Hoagland, still a law school student, during his first term in the Kansas House of Representatives.*

courage to represent your idea in this way in your state legislature, during your first term? Joe took his job as a "representative" of those who elected him to heart. No wonder I was planning to spend the rest of my life with this man.

While we were engaged during the 1974 legislative session, I saw Joe tackle some more big topics that brought him the attention and admiration of his legislative colleagues. He authored the nation's first Comparative Negligence Act. It was not only an important change to Kansas law but a change nationally, as well, to simplify the law and take care of people more fairly. This was a theme all during Joe's legislative career.

Joe's sister, Ruth Ann, had worked part-time at the Capitol during college, and married an attorney in the bill drafting department of the legislature, the Revisor of Statutes. W. Robert (Bob) Alderson, Joe's brother-in-law, played a key role in developing this new statute.

Remember, Joe was still in law school. One of his classes as a first-year student was torts. Unlike criminal law, which involves intentional acts that cause injury to others, torts are broadly speaking unintentional. They are generally accidental or negligent behaviors that cause injury to others.

In that class, he learned about centuries-old English common law doctrines that didn't make sense to him. One was the idea of contributory negligence. That is to say, an injured party can recover nothing from a wrong doer if the injury was in any way partly their own fault. The other concept was called joint and several liability. This meant when there were multiple parties at fault, each was responsible for the entire damage to the injured party. The plaintiff (injured party) in the lawsuit could recover the entire judgment from any one of the defendants, no matter how slight their fault.

This seemed illogical to Joe, so he introduced legislation to change it. In some states the courts had already modified these harsh legal doctrines, but Joe was the first in the nation to change the law by statute. Imagine the confidence that took, from a law student, newly elected to the legislative body.

The idea Joe had was known as comparative negligence or comparative fault. It said that an injured party was not barred from recovering damages so long as they were less negligent than the defendant. It also abolished joint and several liability, providing that defendants are only liable for damages proportionate to their own negligence. This was more fair, common sense.

He introduced a bill to create comparative negligence in Kansas, and it was passed by the House and Senate, signed into law by Governor Bob Docking. It was later adopted verbatim in seventeen other states. In 2023, Joe's accomplishment was admitted into the American Museum of Tort Law in Winstead, Connecticut.

Having written the law, Joe even tried one of the earliest comparative neg-

Speaking at the podium during a House session, Joe carries the comparative negligence bill.

ligence cases, once he was an actual, practicing attorney. As he said during his Oral History Project interview:

"Interestingly, I was able to try one of the first comparative negligence cases as a lawyer, and it was a classic case where comparative negligence was involved. It involved a highway construction worker working on a summer road crew and a person going through the construction zone, speeding . . . and, of course, my client immediately told the police on the police report 'I never saw the car. I never saw what hit me.' Of course, that meant he wasn't looking, and therefore he was negligent, but the person driving the car at a high rate of speed through a construction zone was also negligent. So, we tried that case, and he was able to collect some of his damages as a result."

While Joe was handling these important topics, I had a big problem of my own to solve: How to plan a wedding you could afford when you would need to invite 125 members of the Kansas House and forty Kansas Senators? I simply decided—you don't! You elope, as we did. With the blessing of our immediate family, they gathered around us in August of 1974. After a little party and cutting a cake, they took us to the Kansas City airport for a short wedding trip to Hawaii.

Joe's great-aunt, May Rowland, had arranged for a minister, one of her Unity Church divinity students, to marry us on arrival. It was a sweet, thoughtful ceremony for just the two of us, as we wished. Even our witnesses were simply two ladies from the Unity Church's day care center! This is still an intimate, private memory for us. When we returned to Kansas a week later, we held a cocktail reception to celebrate with our families and 165 of our closest friends!

Our early married life revolved around the political life Joe had created for himself. I was happily thrust into getting to know this whole array of dynamic

Joe Hoagland weds Miss Kansas

Joe Hoagland and the former Ms. Norma Decker were married in an informal ceremony in Honolulu, Hawaii. The ceremony was performed by Reverend Stan Hampson of The Unity Church of Hawaii located at the foot of Diamond Head. Traditional Hawaiian wedding chants and blessings were included in the ceremony. The bride and groom wore traditional Hawaiian wedding leis.

Norma is a senior honors student majoring in Early Childhood Development at the University of Kansas. She is presently employed as a model in the Kansas City area. At K.U. she was selected Best Dressed Coed and in 1970 was Miss Kansas, U.S.A.

Joe is an associate in the law firm of Ross and Wells and serves as a member of the Kansas House of Representatives. He is a graduate of Tulane University and the University of Kansas School of Law.

After the wedding, the couple vacationed on the island of Oahu. They are presently residing in Overland Park.

Our wedding notice from The Squire *magazine, September 12, 1974.*

people involved in political policy to make Kansas a better place. It was a bit daunting for a young, new bride, but I watched the older women and learned. Olivia Bennett, the wife of future Governor Bob Bennett, was a friendly mentor. I soon became the president of the Legislative Wives. It was not called "Legislative Spouses," as yet, because the legislature was nearly all male at that time, with only two women representatives. That would change over time. I truly enjoyed my friendships with these other women who followed their partners to Topeka in early spring for the legislative sessions from January to May of each year.

Again, from the Kansas Oral History Project, Richard Walker and Joe discussed their fifty-year friendship and how they forged a "kind of generational change" in Kansas politics. Excerpts from that interview tell the story of those dynamic years in politics that expanded our worldview and ambitions.

Speaking to Richard, Joe said:

"There were a few young legislators that had been elected before [me], R. H. Miller was elected in 1970. Dave Heinemann was elected I think in '68, but there were very few baby boomers in the state legislature. But in our class in '72, there were a number—Sandy Duncan from Wichita, Neal Whitaker from Wichita, yourself [Richard Walker], me.

Joe's official photo as he became chairman of the House Judiciary Committee, 1979.

"Then the following year, there were more coming along. It kind of became a generational transition because prior to that, all the people that served in the legislature were . . . World War II veterans. There were very few women. I think there were only two in the '72 session when we interned. And that was a big change."

These younger members of the Kansas House learned to organize themselves by forming study groups, following the example of Wendell Lady, another legislator and friend from our home area of Overland Park. Joe recalled:

"Wendell Lady, (a Republican) and Pete Loux, (a Democrat) who was Minority Leader at the time, would meet in the evenings and formulate strategies on how to advance a more urban agenda. Our constituents were very similar. And he (Wendell) had worked, starting when he was elected in '68, to change the formula for the distribution of the state gasoline tax. Instead of distributing based on lane miles to counties, he wanted to have it distributed based on population so that more money from the gasoline tax would go to the cities and less to rural farm-to-market roads. It started the initial clash of rural-urban legislators."

Our home district was now urban, a part of Overland Park, where Joe had grown up, but was now a city. Although Joe and I had already planned to go into agriculture, Joe's constituents were in a suburban area, and he intended to represent them.

Again, from his oral history, Joe recalled:

"When I got into the Legislature with these younger guys, that's kind of what we started to do (study groups.) I remember the first group was R. H. Miller, who had already served one term, you (Richard Walker), and me, and Neal Whitaker from Wichita and Sandy Duncan. And Sandy coined a term for us, for our little meetings as the meetings of the Young Energetic Legislative Leaders Out to Win, which went by the anacronym YELLOW because we talked big, but we were terrified."

YELLOW expanded to include Mike Hayden, who would become Kansas Governor, and Fred Lorentz, who became a district court judge in Fredonia. Robert Frey would become a state senator and majority leader. Bob Whittaker served five terms in Congress. The nine members of YELLOW produced two speakers, two majority leaders, a cabinet secretary, two district court judges, a congressman, a governor, and a lobbyist for the beer wholesalers. As this younger group grew in experience, leadership followed. The baby boomers were taking over.

Again, remember, Joe was still in law school when he first came to the Kansas Legislature in 1972. It was very unusual for a law student to become a member of the Judiciary Committee, but he was placed there by Speaker Pete

McGill. Speaker McGill must have recognized what a serious young man Joe was, as this committee assignment was usually reserved only for attorneys who were already members of the Kansas Bar.

Beginning his second term in the state legislature, we were a newly married couple, commuting constantly back and forth from our district in Overland Park to the State Capitol in Topeka. Joe had passed the Kansas Bar Exam and had joined the Kansas City law firm of Ross & Wells. He was paid $300 per month plus a percentage of the business he could bring in. Harold "Kip" Wells retired as their firm's chief litigation attorney when he got an appointment to be the condemnation attorney for the Alaska Pipeline for the US Department of Interior. As Kip moved to Alaska, Joe took over his cases. Joe's first jury trial was the successful defense of a first-degree murder case. Imagine. After that, he felt unintimidated by any kind of jury trial! In his seven-year career as a practicing attorney, he never lost a jury trial. He won his only case before the Kansas Supreme Court.

Although I had my college degree now and had done some substitute teaching, I hadn't taken a full-time job yet. One part-time job came to me unexpectedly. The phone rang at our little apartment in Overland Park one afternoon. Joe was calling from his law office and asked me, "You can type, right?" I answered yes, sixty words per minute, no errors. (I'd taken a typing class in high school.) He replied, "I just came back into my office, and my secretary was eating her lunch with her feet up on my desk. I fired her. Can you come be my secretary for a while?"

This was another learning experience for me. Mrs. Margaret Hathaway was the office manager in charge of the secretarial pool at Ross and Wells law firm. I don't think she was delighted when Joe informed her his young wife was coming in to take over his secretarial duties! This was before computers, before photocopy machines, back in time even before typewriters that could be programmed. We're talking about typing on onion skin paper, five copies at once, with carbon paper between each! Does anyone even know what carbon paper is now?

I learned so much, not only how to get along with a prickly older woman who was accustomed to being in charge, but how to understand, spell, and phrase legal documents, drink coffee, and take my place in the "pecking order" of an established office. I learned the lifelong skills of restraint, diplomacy, and patience. And I learned how to get along with my husband in a formal office setting, as his assistant, a most valuable skill for the rest of our lives together. Sometimes I think I've never stopped being his assistant, but then he's also been mine throughout our years together. These were skills we used to work together to organize the Black Hereford Association. That's teamwork.

Joe and I had agreed I would do part-time work while he was spending months at a time in Topeka. We wanted to be together, so I needed to be able to move back and forth with him. We were young and had no children yet, so this was possible. We had already seen divorces happen when couples weren't able to be together for those legislative months in Topeka. With one person in Topeka and one back in the home district, marriages seemed to fall apart. We were determined not to let this happen to us.

So, for part-time work, I began to do a few modeling jobs, runway fashion shows for Swansons and Halls on the County Club Plaza in Kansas City. Swansons was such an elegant store. They would have "trunk shows" where several young women would model sample garments (all in one small size) sent to the store by Chanel or Yves Saint Laurent. Wealthy local ladies would come, drink champagne, and select and order their choices after seeing us model the clothes. These were exquisite garments in gorgeous fabrics—fun to wear even for just a little while!

Television commercials were my next line of part-time work. First was Racket Merchandise Company, doing television commercials for my father-in-law, Virgil Hoagland. They were great fun and gave me a special relationship with Virgil, whom I greatly admired.

I had learned to know my father-in-law by doing some secretarial work for him, too. His mother, Berniece, had done most of the typing of letters and documents for Racket, but she was reaching an age when coming downtown was harder for her. So, he asked me to step in. First, I found Berniece had typed on a manual typewriter, like you'd see in a museum! After a couple of days, I talked to Virgil and explained I couldn't work as efficiently without an electric typewriter. Now, Virgil wasn't known for spending money unless it was really, truly necessary. I can still see him pacing around his office, smoking his pipe, and jingling his keys in his pocket, as he thought over this expense. He relented and a new electric typewriter was installed in the office. This meant a lot to me. I continued to do any work needed for him, and for Racket.

So, when the idea of me doing TV commercials came up, Virgil and I already had a great working relationship. We did advertising with focus on the historic story of Racket, making me feel a true part of the family. Bless Virgil: when he suggested his daughter-in-law for doing commercials to his advertising company's agent, Sam Bradley, Sam's first response was a horrified, "Oh, no!" But Sam found I had good timing and presence, and we worked together well over several years on Racket's commercials.

This led to more advertising work as a spokesperson for several nationwide companies: MasterCard and Holiday Inn. In the 1970s, Holiday Inn had a creative new idea—the Holidome! This was a hotel design centered around a

The television studio of WDAF, Channel 4 in Kansas City, Missouri, where Norma filmed many of her TV commercials.

large, domed area with a swimming pool, a very new idea—and I got to advertise them. I was given the name Donna Desmond when I was a spokesperson for MasterCard. My real name Norma Hoagland just didn't roll off the tongue!

I had one account that became a regular weekly commercial. Based in Denver, Colorado, there was a grocery chain named SuperFoods, and I played "Donna" from SuperFoods, announcing the weekly specials. One winter when we went skiing in Colorado, people even recognized me! Great fun! It was a perfect, stable account for me, with Joe proudly telling people I was making more money on one afternoon a week than he was making all week as an attorney!

Some television studios were all union, so I had to join the American Fed-

MasterCard ad, featuring Norma as "Donna Desmond."

eration of Radio and Television Artists. We were Republicans, but as a union member, I helped Joe get endorsed by the AFL-CIO. Very few politicians in Kansas have ever been endorsed by the Association of Commerce and Industry and the AFL-CIO, making it almost impossible to beat Joe in a general election. But then, his elections were never close.

There were other big issues Joe was involved in during those twelve years in Kansas politics, fascinating for me to watch. I would go to the Capitol and listen to the debates from the upper floor gallery.

Government reorganization took place during Joe's years, changing an outdated government of independent boards and commissions to a cabinet form of government, under the Governor's control. It began with the Department of Revenue, then over the years included Social and Rehabilitation Services, Health and Environment, Agriculture, etc. Joe authored the bill to establish the first state agency for Environmental Protection.

During the creation of the Department of Transportation, Joe helped a whistleblower be heard while he was vice-chairman of the Governmental Organization Committee. The whistleblower came to Joe with information about irregularities in the construction of Highway 59 south of Lawrence. That dust-up included Curt Schneider, then the attorney general of Kansas, who seized logbooks needed by the whistleblower to prove his claims. The press followed this closely, one of several issues that put Joe on television and in the press

Claims Records Removed In Highway Investigation

By Ray Morgan
Kansas Correspondent

Topeka—State Rep. Joe Hoagland (R-Overland Park) said last night he has sent a letter to L. W. Newcomer, Kansas highway director, requesting copies of highway construction records in the Lawrence area after alleging the records disappeared from the district office there.

Hoagland is a member of the House Government Organization Committee which heard testimony from a Kansas Highway Department inspector alleging irregularities in Kansas highway construction projects.

Curt Schneider, Kansas attorney general and a Democrat, said last night after two days of questioning of John W. Young, the highway inspector, that Young had failed to produce any substantial proof of irregularities in highway operations.

Schneider also sent a letter to Gov. Robert F. Bennett, a Republican, asking Bennett to provide any information he may have of any alleged irregularities in the conduct of Kansas highway operations.

Young declined to make any comment on the questioning by Schneider at the end of the hearings on the advice of his attorney.

Hoagland, whom Schneider has indicated may be subpoenaed to outline what he knows about the matter, said he had sent the letter to Newcomer because he believes the papers he seeks confirm the charges made by Young.

Sources involved in the investigation have indicated that when Young returned from testifying before a legislative investigating committee last Friday, the log book of the project to which he referred, Log Book 40 C, had been taken from the Lawrence office by a state highway official from Topeka and Young's department supervisor.

They also say the log book was in Schneider's possession during the first day of the inquisition on Wednesday when he was questioning Young.

Another volume which Young indicated would help substantiate his allegations, Log Book 70 A, was requested by Young from his supervisor but he was told that the volume was not available. It also showed up at the questioning conducted by Schneider yesterday.

Hoagland said he wants to see the records because, "Schneider has stated that Young's allegations are unsubstantiated, yet if these records contain what Young claims they contain, then there would exist, in my opinion, a substantial basis for an extensive investigation."

Hoagland also alleged that Schneider is attempting to cover up the facts in the highway department under Robert Docking, former Democratic governor, because members of the Kansas Highway Commission, contractors and suppliers and former highway department officials made large contributions to Schneider's 1974 election campaign.

A check of a list of contributors to Schneider's November general election campaign in the office of the secretary of state showed more than $14,090 in contributions by either highway contractors or department officials.

"It would be improper at this time for anyone to suggest that Schneider is not acting in good faith simply because his wife is making nearly $10,000 a year as a secretary to the assistant director of highway administration," Hoagland said.

In his remarks before a legislative committee during hearings on a Senate bill to create a department of transportation under a secretary, abolishing the present Kansas Highway Commission, Young alleged his orders were often overriden in connection with highway construction projects.

Front page article of the Kansas City Times, *as reported by Ray Morgan, Kansas Correspondent. "Claims Records Removed in Highway Investigation" was one of the articles concerning the whistleblower Joe helped be heard, despite Attorney General Curt Schneider's obstruction.*

during our years in the Kansas House of Representatives. Here was another skill, managing the press, that Joe would use later, founding the Black Hereford cattle breed.

The tone of politics was so different in the 1970s than what we experience today in 2026. There were differences of opinion, naturally, about issues. The lines of difference tended to be more urban vs. rural, rather than Republican vs. Democrat. We all respected each other's responsibility to represent our own group of constituents. We could disagree and still be civil and friendly. Some young couples even shared apartments during the legislative sessions to save money. We counted among our good friends those from both political parties. One wonderful memory I have of those years is dinner with Rep. Jim and Linda Slattery of Topeka, in their home. After dinner, we listened to recordings of John F. Kennedy's speeches. Jim, a Democrat, would go on to Washington and serve six terms in Congress. Yes, we were all young and idealistic, but pragmatic and respectful of the job of governing.

CHAPTER THREE

POLITICAL LEADERSHIP AND CATTLE

In 1977, Joe found himself in the Minority Party following the election of Jimmy Carter and the backlash directed at Republicans after the Watergate scandal. Despite this, Joe was able to get a seat on the powerful Ways and Means Committee which handled appropriations and emergency legislation.

Toward the end of that 1977 session of the legislature, an interesting and noteworthy topic came up. Looking back, we can scarcely believe there could be opposition to such an idea.

Southwestern Bell Telephone Company developed a new product called 9-1-1 for emergencies. Legislative approval was required to authorize its use. The Greater Kansas City Metro area was particularly complicated because the state line went right through the middle of the city. There were two different counties on the Missouri side and three on the Kansas side plus dozens of different municipalities, and each of these parties involved required approval.

In Missouri, Senator Harry Wiggins from Kansas City introduced the enabling legislation there. Joe Hoagland of Overland Park, on the Kansas side, introduced a comparable bill, it was noted in the press. The legislation would have to pass in both states in substantially identical form and take effect at the same time. This wasn't going to be easy.

There was immediate opposition from these local governments who did not want to lose control of their own police and fire dispatch. A local Kansas City newspaper, *The Sun,* had a headline on March 25, 1977, that read: "9-1-1 STALLS IN TWO STATES."

With only one week remaining in the 1977 legislative session, Wiggins had changed his mind after reviewing the bill's implications and removed it from a special accelerated Senate Calendar reserved for non-controversial bills. Joe presented the bill to the Kansas delegation from the KC Metro area. Press clippings from the time quote him as saying, "A bill should at least be introduced this session to signal the Missouri Assembly that Kansas is seriously working on the telephone system." Joe wasn't about to let this life-saving measure be dropped. He got it introduced into the Ways and Means Committee. Then an interim committee was set up to work on the issue so that it was ready for debate, and hopefully passage, during the 1978 session.

Remember, this occurred at a time when there were long-distance phone calls. The telephone company charged more for individual phone calls outside your own community. An easy to dial emergency phone number was no doubt inevitable. But in 1978, it seemed impossible in the Greater Kansas City area where local calls crossed those state lines between Kansas and Missouri.

From Joe's Kansas Oral History Project interview again, he remembers this important piece of legislation.

"One of the things that was interesting that we take for granted today was the 911 emergency phone system. That was before the deregulation of the telephone industry. It was Bell Telephone Company, which was regulated by the corporation commission for telephone rates and everything else. And to be able to set up an emergency phone number, the same as the information number,

Joe at the podium on the floor of the House, speaking in favor of the 911 emergency phone number authorization bill.

which was 411, they needed governmental approval to be able to do that, to offer that product. So, they came up with this idea of an emergency phone number that people could dial that would be so simple that even a four-year-old could dial 911, if it was necessary, and there have been stories of that actually happening.

"These municipalities did not want to give up their control over the dispatch of their fire department and their police department, and they opposed (the 911 emergency phone number.) The League of Municipalities . . . was in opposition, they had too many urban cities that didn't want to have that central dispatch. We got it through. It passed, and I remember Ed Schaub coming to me afterward. He was a magnificent lobbyist (for Southwestern Bell Telephone Company). . . . It was more his work than my skill on the floor debate in getting it through . . . and of course, 911 is available everywhere today.

"But after it was done, I remember him coming to me and saying, 'Now, we (the telephone company) have another project, and we need to be able to set up towers to collect wireless transmission of phone calls and move those phone conversations from one tower to another, and we need to be able to acquire easements on private property to set up these towers. And we would use this system to operate in cells, transmitting telephone calls from one cell to the next,' and he wanted me to carry that bill (on the floor of the Kansas House.)

"I said, 'That is the craziest idea, Ed. What are you going to call them?' 'Well, they're going to be called cellphones.' . . . But I said, 'No, I've done enough heavy lifting for you. I'm not going to do any more for the telephone company.'"

The rest is "cellphone" history!

The greater metropolitan area of Kansas City was the most difficult place to establish the 9-1-1 emergency phone system in the entire country. Not only were there numerous municipalities, but there was also a state line between two states that ran through the middle of the metroplex.

Even after the enabling legislation passed in both states allowing the establishment of 9-1-1 emergency calls, it was 1983 before the first call was made. It united more than one hundred area police, fire, and emergency medical numbers under the easy-to-remember three-digit number in sixty-five cities—in six counties—in two different states.

Before his election to the Missouri State Senate, Harry Wiggins had been the executive of Jackson County, Missouri, and was well respected for pushing through the construction of the Truman Sports Complex, home of the Kansas City Royals and Chiefs. With his help in Missouri, and Joe on the Kansas side, both State Legislatures approved the 9-1-1 emergency phone number system for both states. The legislation brought both men to statewide prominence after its passage in 1978.

Harry continued to serve in the Missouri Senate until 2003. He said the creation of the 9-1-1 emergency system was the greatest accomplishment in his career because it saved so many lives and cost very little money to accomplish. Harry was elected Kansas City's first senate majority leader on the Missouri side. He was majority leader from 1980 to 1984. Joe was elected majority leader on the Kansas side at the age of thirty-two, making him the youngest person in Kansas history to serve in that position. Joe would serve from 1983 until 1985 in that same majority leader position in the Kansas House of Representatives. They remained good friends until Harry's death in 2004.

In 1979, Joe began his third term in office. Leadership positions would be coming for Joe and this new and younger generation of leaders in Kansas.

As I've mentioned before, most of the members of the Kansas House were WWII veterans, years older than us. When Joe was elected majority leader, he was only thirty-two. He has always given me some credit for that one-vote victory to elect him majority leader, although I'm not sure it's justified. Since I was president of the Legislative Wives, I did make sure all the ladies knew Joe was running for the leadership position. I always felt it was his excellent skill in handling policy and debate in committees and on the floor of the House that gave him those votes, not my gentle suggestions to the ladies.

As majority leader, 1983-84, Joe managed the schedule for floor debate on bills. His desk was beside the podium at the Speaker's right hand.

By the time Joe was majority leader in the Kansas Legislature, he was also the managing partner of his law firm. The firm was now Ross & Hoagland. They had two locations and a team of lawyers and clerical staff to manage. At this time, Joe was admitted to practice before the United States Supreme Court.

We were ready to invest in a place to live during the legislative sessions, closer to Topeka, while still maintaining a house in our legislative district. Buying some farmland was our first choice. Our families both owned lands, mainly farms and crop ground. My father had always taught me that land is the only solid investment, the only true wealth. Having grown up on a dairy farm, I understood what the constant daily work of a dairy means. So, when our "negotiations" as a married couple began about what kind of agriculture we would continue in the family tradition, I firmly said "not a dairy." So, raising beef cattle we agreed on, a type of agriculture we could manage around our busy political life.

Our farm outside Holton, north of Topeka, was small. We inspected and bought it on the first car ride for our first-born daughter, Anne, when she was only two weeks old. There was a small farmhouse with only four rooms and one bath. It was a bit primitive; with a cistern we had to haul water to fill and only a

Joe, Norma, and our infant daughter Anne, with Joe's father, Virgil, at our farm near Holton, Kansas.

stone-walled basement to flee to in case of a Kansas tornado. This was important. We used that basement several times in our years there, once—too close a call—when a barn across the road was destroyed by a spring tornado! But it was a snug little home, with a nice fireplace and big windows. It had adequate barns and we had enough land to raise hay for our cattle. The pastures were brome and fescue with nice ponds. As we added machinery to our place, we began to do the crops ourselves. First, we had a couple of steers for beef, then we were ready for mother cows and to begin a cow-calf operation. This began our first search for what breed of cattle we wanted to raise.

One of Joe's law partners was Bob Elliott, whose brother-in-law was Steve Renko, a professional baseball player. Steve had been the quarterback at Kansas University when Gale Sayers, future Chicago Bear, played football there. Steve had a great arm and was drafted into the NFL and also Major League Baseball. Joe's law firm negotiated his contract to play baseball instead of football. He played fourteen seasons in the big leagues. He was the only person to have both thrown a touchdown pass to Gale Sayers and strike out Hank Aaron. As a sports fan, Joe was impressed by Steve, but it was his brother-in-law Bob who guided a most important step in our cattle history.

It was Bob Elliott who introduced us to Dewey Rounds, communications director of the American Hereford Association. Dewey convinced us to raise registered Herefords. They were the nation's most popular breed of cattle at the time. They were known for docility and feed efficiency. For me, docility was primary, knowing I would be the initial hired hand. My father had always had a few Herefords around our dairy farm to provide beef for the family. So, I knew the gentle temperament of Herefords. Joe could easily commute to Topeka from our farm, and I did the lion's share of farm work. We bought a small Ford 8N tractor so I could more easily feed cattle and do chores on my own.

As we settled into our city vs. country life, I found local girls to babysit our first daughter Anne and later our son Dirck, so that I could still attend some daytime House sessions to listen to the policy debates and votes. Joe always alerted me to any interesting topics on the House calendar. And we both enjoyed the evening gatherings at the Capitol. Many times, I would be in a cocktail dress with my rubber chore boots on, doing last-minute checks on cattle before driving in for an evening's fun!

Cows and politics—what more could an ambitious farm girl hope for? These early years of political involvement made us realize so clearly how politics impacts everyone. We have stayed informed about issues from those early years onward, discussing politics every day, most often over coffee together every morning. That is our permanent daily ritual—morning coffee together.

At about this same time, Joe was appointed to the Kansas Judicial Council, an advisory committee to the Supreme Court. It was always chaired by a justice next in line to become chief justice of the Kansas Supreme Court. I knew this was another recognition of Joe's work ethic.

Not every issue had a positive ending to it, however, for Joe politically, and for us personally. From Joe's interview for the Kansas Oral History Project:

"Here we are, fifty years later, talking about an issue that was a problem fifty years ago. Gun violence was an issue before the legislature in the 1970s. I authored a couple of pieces of legislation with some Democrats from Wichita; Gene Anderson was one of them, Paul Feliciano, on the other, to make it a bipartisan bill.

"One . . . if you are going to use a firearm in the commission of a felony, you're going to face an automatic prison sentence. It was a mandatory prison sentence for crimes committed with a firearm. It is still on the books today. . . .

"Then Gene Anderson and I authored a bill to ban the sale of handguns with a barrel length of less than twelve inches, thinking that handguns are not used for hunting, and we could help eliminate gun violence if we just took these handguns off the street. Well, the first reaction was just the sheer introduction of the bill resulted in a run on handgun purchases. And it absolutely got nowhere. It didn't even have a committee hearing. . . . Maybe we were just trying to make a point that we've got to do something about it [gun violence] . . . but here we are, fifty years later, we're still trying to find a solution to that. . . . I got sued by the Kansas Chapter of the NRA."

Okay, Joe admits banning the sale of handguns was not a good idea. But the mandatory prison sentence for crimes committed with a firearm has been very successful, prosecutors have told Joe. It has dramatically reduced the number of jury trials in aggravated robbery and assault cases. As it was explained to me, this is because the law was written to say "use of a firearm" must be an element of the crime, plea negotiations became much easier.

To further explain: When a defendant knows he is guilty, why not have a trial and see if he can get off? Before this change in the law, defendants had nothing to lose by going to trial. With this new law, if they lost their trial, the judge would have no option but to send the defendant to jail. Prosecutors could decide if the gun charge was in or out of the case. It gave prosecutors leverage over the defendants to plead guilty (if they were) in order to avoid going to jail, if the gun was left out of the official charge.

The new mandatory prison sentence law has been very successful. Banning the sale of handguns was wildly unsuccessful. Americans like their guns.

Yes, we were a young and idealistic couple, trying to do our best to represent and protect vulnerable citizens of Kansas and we got sued by the NRA. I'd never been sued before. I still remember the stunning and frightening impact that had on me, as my name was included in the lawsuit as well as Joe's. Politics, as well as agriculture, is a team sport, and often not a friendly sport. This all was preparation for the next phases of our life together.

Reflecting on his role in the passage of all these important changes to our laws, Joe said, "I always found it awkward as an attorney arguing a case before a judge involving an act of negligence, a criminal case involving a firearm, or a small claim. If a judge disagreed with my argument I was tempted to say, 'Well, I wrote that law!' but I never did." (Thankfully.)

During his fourth term, Joe was appointed to the National Commission on Uniform State Laws. He was named by the Kansas Bar Association for this honor. This was no doubt partly due to his landmark statutory change to tort law, comparative negligence. Attorneys from each state would meet one week each year to draft legislation for passage by the various states where the state laws should be the same in each. Examples of these would be the Uniform Commercial Code governing business transactions and the Uniform Rules of Evidence for courts. This was profoundly a states' rights organization designed to prevent the federal government from taking over the power that states held to govern themselves. Yet the states were part of one nation where the rules needed to be the same for everyone. This was an honor for Joe, and he enjoyed the policy work.

The attorneys on the commission were some of America's finest, selected by the Bar Association in each state. Illinois had sent Albert Jenner of Jenner & Block of Chicago. He represented Richard Nixon during the Watergate hearings. Joe was with James Bush of Arizona when James got the call that President Reagan had named his former law partner, Sandra Day O'Connor, to the US Supreme Court. From Colorado, Ann Gorsuch was Joe's seatmate during the meetings. She was in-house council for Coors Beer. President Regan later appointed her to head the Environmental Protection Agency. Her son, Neil Gorsuch, was later appointed to the US Supreme Court by President Trump. We remember him as a youngster, attending these meetings with his family, as we attended with our young family.

Having never lost an election or a case before the Kansas Supreme Court, Joe was now making more money than his father was running the family business. More importantly, Joe had gained his father's respect by making his own path successfully, after their parting over Joe attending law school. That rift had long been healed, and both of his parents were proud of the work Joe had done as a politician. Virgil doubted Joe would even be interested in the family business, but Virgil had decided to retire. He announced he would either sell the company or give it to his children, Joe and his sister Ruth Ann.

Joe decided at this point to retire from law and politics and manage the

Celebrating their fiftieth wedding anniversary in 1994, Joe's parents, Ruth and Virgil, are pictured along with Joe and his sister Ruth Ann Alderson.

Consolidating our farming operations,1985, in Leavenworth County, Kansas, we were breeding traditional Horned Herefords and raising crops on the bottomland along the Missouri River.

Our family is complete! Joe and Norma with son Dirck, daughters Anne and baby Emma.

family business from that point on. We prepared for a new phase of life. We began to consolidate our cattle and farming operations to our Leavenworth ranch, along the Missouri River. There was room to grow our herd, and it was a relatively easy commute for Joe to the Racket offices in downtown Kansas City. It would be perfect for both of us to do the work we wanted to do. Partnership.

Our daughter, Anne, was already in public schools by this time and had an odd routine. She attended one school every fall semester, in our home legislative district in Overland Park. Then she would attend a different school for the spring semester from our farm, near the State Capitol, during the legislative session. We had sacrificed her first years of education to our wish to be together. As she changed schools each year, Anne has always jokingly said she missed learning subtraction completely in school! Our son Dirck was about to begin school. This was a time of growth in so many ways for Joe—and me. We had added another lively little daughter, Emma, to the family.

Our children were growing up and needed a more settled life. We couldn't continue to sacrifice their education for our partnership. Joe's decision to leave politics came at a good time for our young family.

CHAPTER FOUR

A LEG UP—FAMILY HERITAGE

Joe did not start from scratch. His family had generational wealth and social connections not only in Kansas City but through his cousins in New York. The Hoaglands were accustomed to being involved in their community. His father, Virgil, was president of the school board. His grandfather, Daniel, was a charter member and principal benefactor of a new age religion. These new religions were popular at the turn of the last century.

On our fiftieth wedding anniversary, 2024, we celebrated with our family at Unity Village, Missouri, an important place in our family's history.

This new age religion was called the Unity School of Christianity, headquartered in Unity Village, Missouri. They did not believe in original sin and believed that Jesus and Mohammed were both great teachers and simply different paths to the same end. They believed in reincarnation. That's probably why they were all vegetarians, Joe says. In Unity, salvation was unnecessary, as no one needed to be "saved," as Joe's Great-Aunt May would say. They tried to explain all religions as being one. Hence, the name Unity. Remember, Daniel's father, George, had been a minister in Topeka, Kansas, after the Civil War. This was a kind, gentle, and spiritual side of the family.

The other side of the Hoagland family were meat-eaters. They, led by Sylvester, were merchants in the old Dutch tradition of buying and selling. He founded the Racket Merchandise Company in 1891. That company still exists today, 135 years later, under the leadership of the sixth generation of Hoaglands. It has always been a source of wealth for the family to buy land, from cattle ranches in Colorado and Kansas, citrus orchards in the Rio Grande Valley to mining and timber near Yellowknife, Canada, in the Arctic Circle.

Racket Merchandise Company started off in 1891 across the street from the old city hall in Kansas City, nearly on the banks of the Missouri River. It was an outfitter for pioneers traveling west to join the Santa Fe Trail or the Oregon Trail. Located right in the River Quai area, these adventuresome travelers would gather, often getting off steamships and railroads. Racket Merchandise Co. sold general merchandise such as cast-iron skillets, chains, bullwhips, anything and everything to fill your covered wagon to go west.

As the years passed, Racket Merchandise Company changed with the times. Racket sold to a more urban customer, selling housewares along with hardware. After World War I, it supplied kitchen equipment to restaurants and hotels throughout the Midwest.

Several of the founding Hoagland family had no children or heirs to take on the management of Racket after Sylvester Hoagland. Ownership centralized in the Julian Hoagland family. Julian and Berniece Eddy Hoagland had two sons, Julian and Virgil, whom the family assumed would follow in the family business.

Tragedy struck when Virgil's older brother, Julian, died from drowning while trying to save a young girl's life. She survived; he did not. That left Virgil, at a very young age, to learn to run the business from his father, Julian. As his father's health soon declined, Virgil and his mother, Berniece, worked hard together, as a team, to keep the family business afloat in the 1930s as the Great Depression hit America. Dedication and hard work became Virgil's life—a legacy he passed on to Joe. And the example of their teamwork wasn't lost on Joe, either.

Joe's father, Virgil, became the general manager of the family business, Racket Merchandise Company, in 1937. He is pictured on the far right with the Porretta brothers, who were career employees.

Virgil was a merchant at heart. He is credited with not only stabilizing Racket during that hard time of the Great Depression but also moving it into a new age. Air travel was just beginning in the 1930s when two airlines started up in Kansas City: Trans World Airlines (TWA) and Braniff. Virgil worked to sell products to the airline industry—a new business, but again taking care of travelers, you might say.

TWA became the first airline customer of Racket when they started passenger service in Kansas City. The early airport for Kansas City was near the legendary stockyards in what was called The West Bottoms. According to family legend, flies from the stockyards would flock into TWA planes when they landed and opened their doors. As a local business, Virgil's first contract with TWA was to sell fly swatters!

From that humble beginning, Virgil moved into many products used by customers on TWA's planes. Racket became the principal supplier of passenger service items such as chinaware, glassware, linens, blankets, and stereo headsets for inflight entertainment to TWA. As time went on, he developed contracts with many other domestic and international airlines. After the passing of Berniece in the 1970s, Virgil alone owned and operated Racket Merchandise Company. He was the fourth generation of Hoaglands to do so. The airline supply business was fully half of the company's revenue in those days.

Ever since 1891, Racket had been open to the public and by the 1970s, Virgil had five retail hardware stores in the Kansas City area. Joe still remembers his high school days when he and his sister, Ruth Ann, spent weekends and summers working as salesclerks in those stores. He remembers one day in 1963 at the store at 713 Walnut in Kansas City, selling former President Harry Truman a new electric toaster. At the store located at 83rd and Mission Road in Prairie Village, he often waited on Hank Stram, then the head coach of the Kansas City Chiefs, who lived nearby.

But by the end of the 1970s, big box stores had moved into the hardware and housewares market. These large national brands squeezed out the local independent merchants and Racket was no exception. Virgil had bought farmland over the years, mostly as investments. He started selling the stores and putting that money into citrus groves in South Texas. Virgil and Ruth began to spend their winters in Texas where they owned a house on West Pecan Boulevard in McAllen. Virgil sold his last hardware store in 1980. To avoid paying too much in taxes, he invested more in citrus orchards. Texas is famous for its Ruby Red grapefruit. He bought blocks of orange and grapefruit trees, ranging in size from 5-acre to 30-acre tracts in Hidalgo County. By the time Joe had left politics to run the family businesses, Virgil was one of the ten largest citrus producers in the Rio Grande Valley.

However, growing fruit was never as profitable as selling orchards to become housing developments. It became such a large business with over twenty different properties to manage that Virgil hired a full-time manager, Billy Bob Bates, who hired a young Texas A&M graduate named Mark Fryer. Virgil and Ruth became Texas residents, living in McAllen nearly year-round.

By the mid 1990s, Joe spun off the citrus business from Racket and created Lone Star Citrus LP. It is still in operation today and has been another source of family wealth in recent years.

After the terrorist attack on America, September 11, 2001, the airline business changed. Long security lines meant travelers arrived early and ate meals while waiting in airports instead of planning on eating meals in flight. This lessened Racket's airline business quickly. In 2008, Joe sold the largest orchard owned by Lone Star to an investment group in Switzerland. With those proceeds, he bought a ranch that adjoined one of ours in the Kansas Flint Hills. In 2024, Joe sold a 14-acre grove his father had bought in 1978, which was now positioned on a 4-lane highway. That land would become a hospital, and the money was used to buy a ranch in Colorado, near Fairplay.

In 2002, we visited Ruth and Virgil in McAllen, Texas, to visit the citrus groves. From left to right: Brandon Bauer, Anne Hoagland Bauer, Emma, Norma, Dirck, Ruth, Joe, and Virgil.

But returning to 1989, Joe had positioned Racket to only do business with commercial airlines. They sold something to every airline in America, most airlines in Europe, and even some airlines in the Middle East and Asia.

One of Joe's customers was Eastern Airlines, based in Miami, Florida. Eastern also owned a shuttle service between three busy hubs: Boston's Logan Airport, Washington National, and New York's LaGuardia Airport. Flights were every thirty minutes, operating seventeen Boeing 727 Aircrafts. Eastern was about the last carrier to use this aging aircraft. It had three engines in the tail of the plane and needed a three-person crew of pilot, copilot, and navigator to fly. Costly and inefficient compared to newer aircraft like the Boeing 737 or McDonald Douglas's MD80, Eastern Airlines decided to sell their shuttle operation to a New York developer named Donald Trump. Trump owned apartment buildings, but this was his first foray into the airline business. The flights were short, so there was no meal service, but they had pillows, blankets, and used ice buckets and coffee pots. All were things that Racket sold.

Joe met Donald Trump in New York in 1989 after Trump's purchase of the Eastern Airlines Shuttle service. It was not an airline per se but a series of landing slots and aircraft at those three busy airports. It had no administrative offices.

The Trump Shuttle operated from 1989 to 1992, flying the aging and fuel inefficient Boeing 727s.

Trump called a meeting with Joe and all the other companies that had been suppliers to the Eastern Shuttle. Joe said he was surprised at how tall Trump was and that he really stood out in a room full of people.

As the new owner of the Shuttle, Trump announced he would buy three years' worth of inventory rather than set up a purchasing department for his new airline. All our fellow suppliers thought this was odd, but we all reluctantly agreed.

Trump didn't pay any of us. Lawsuits followed, which forced the Trump Shuttle into bankruptcy. It operated from 1989 to 1992, during those three years. This was the first of several bankruptcies for Donald Trump who would later become president of the United States.

Coffee pot produced by Racket Merchandise Company, under contract with Trump Shuttle. (Photo by morgan jane photography, Leavenworth, Kansas)

In 1992, Racket purchased a company in California owned by Arthur Michaud, who had been Tommy Dorsey's band manager. (Think of 1940s Big Bands.) Arthur had been the first person to import Gucci products to the United States. He used these Gucci bags to make overnight amenity kits for Pan American Airlines, a customer of Michaud. The acquisition of Michaud by Racket was strategic and expanded Racket's customer and supplier base. We hadn't done amenity kits very successfully before this.

Michaud also sold amenities to hotels and cruise ships. Arthur sold ice buckets, slippers, and shower caps to Caesar's Palace, bath robes to the Four Seasons Hotel chain, and little bottles of shampoo and body lotion to Carnival Cruise Lines. This would be a great expansion for us.

As the negotiations for this acquisition were going on, Joe and I were invited to Aurthur Michaud's eightieth birthday party. We sat at the head table with Arthur and Gladys, his wife. Also at this table was Rudy Vallee, an aging but still famous film star and his wife, Eleanor, draped in emeralds. My mother was so impressed that I danced with Rudy Vallee! We must have made a good impression because shortly thereafter Arthur agreed to sell us his company, Michaud Inc.

One of the employees that came to us with Michaud was a colorful man named Jerry Thomas. He had recently retired as vice president of international sales for Capitol Records. In the early days of The Beatles, Jerry had traveled with them on their world tour and knew each of them well. Jerry's main account with Michaud, and now Racket, was with Delta Airlines because of his connections in the American South. He graduated from Vanderbilt University and moved to California so he could be within an hour's drive of the ski slopes or the beach. He told Joe once that after forty years, he had only gone skiing three times and surfing twice. Jerry was a great storyteller, and his life gave him good material to work with. He was a born salesman—our customers loved him.

Joe employed three full-time designers in the Kansas City office: one from Japan, Yoshi Sato; one from Italy, Christina Serafini; and one American, Andy Harris. We built manufacturing facilities in Kansas to make pillows and blankets for airline use. A cosmetics plant was built in Topeka, Kansas, to make Racket's own moisturizing lotion for long flights in the dry air of a pressurized airplane cabin.

With hot meals served on most flights over one hour in length, reusable chinaware and flatware were common. More meals were served on airplanes than

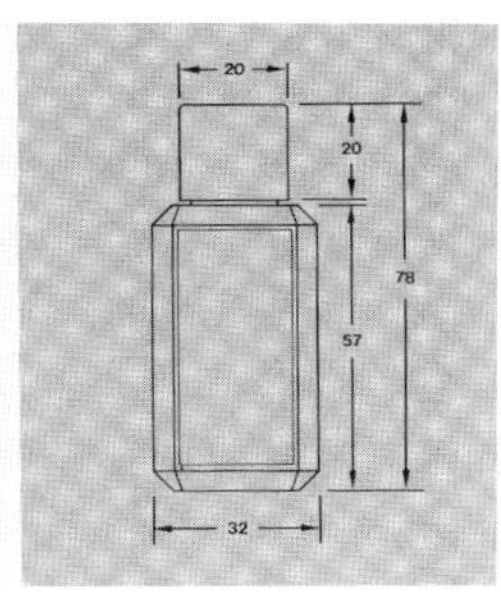

Detail of Michaud line of amenities on right. Yoshi Sato, head designer at Racket on the left.

anywhere else except for the US Army and McDonalds. Even in economy class, meals were served on a chinaware dish with a small sized set of stainless-steel cutlery. The quantities used by the airline food catering industry were huge. Each piece was washed and reused before another flight. Joe would say he made a small fortune as a "teaspoon salesman." He sold three million stainless-steel teaspoons in one year, just to United Airlines alone. To put that in perspective, three million teaspoons weigh the same as eight Ford F-250 pickups! We had dozens of flatware contracts with airlines all over the world.

The 1990s were an extremely busy and lucrative decade for us. Joe was traveling much of the time, in the prime of his business years. We missed him at home, but these were such important years for the company—and for filling Joe's "toolbox" of experiences he would need in the future. He was always making contacts and thinking ahead to our future.

Adding Michaud to our line of products gave us a full "menu" of passenger service items to sell. But Joe found increasingly fewer items could be made domestically in the United States.

Racket's Senior Management Team in 1997, from left to right: Amy Pittman, Hotel Sales; Mark Fryer, Citrus Operations; Tim Cook, Airline Sales; Yoshi Sato, Head Designer; Jerry Thomas, Airline Sales; Joe Hoagland, President and CEO; John Dulleson, Cruise Line Sales; John Postlewait, Executive Vice President.

Success Stories

64

JOSEPH HOAGLAND
Racket Merchandise Company

Fast growth sometimes requires a little luck. Joseph Hoagland says Racket Merchandise Company, which supplies passenger service products to airlines, has been on the right side of several recent airline mergers.

Joseph Hoagland runs what may be the most misunderstood business in the Corporate Report 100. Its name, Racket Merchandise Company, suggests a racquet-sport manufacturer with customers like John McEnroe and Chris Evert Lloyd.

But the family-owned, Kansas City, Missouri, company has nothing to do with tennis or any other sport. It has customers like Carl Icahn and Frank Lorenzo and supplies passenger-service products such as meal trays, stainless-steel flatware, slippers, headsets, and wine baskets to commercial airlines.

The company's 1986 sales of $7.1 million, an increase of 373 percent since 1981, has earned it a place in the Corporate Report 100 for the second year in a row. "It'll be interesting to see if we make it another year," says Hoagland, the company president, acknowledging the difficulty of sustaining rapid growth. But he boasts, "We're on our way to $10 million" in sales for 1987.

Racket Merchandise has grown to be one of the largest among six or seven companies in the United States and Canada that supply the airlines. Shortly after airline deregulation, Racket hit its peak in terms of customers, selling to about 120 airlines, Hoagland says. Consolidation in the industry has trimmed the number to forty customers, one-fourth of which are based outside the United States.

Since 1984, Hoagland has played a bigger role in the management of Racket, and that, along with airline mergers, has helped the company grow. A resident of rural Leavenworth, Kansas, Hoagland served in the Kansas Legislature for twelve years, including two years as House majority leader, before retiring from politics in 1984. "I came to the point where I had to decide between full-time politics and full-time business," he says, noting that politics limited his involvement in Racket to nine months a year and three days a week. "I looked fairly seriously at running for Congress, but I just wasn't ready for that kind of commitment.

"I'm very content with doing what I'm doing. It's fast-track, glamorous, international travel — a lifestyle like politics in a lot of ways."

Racket Merchandise Company has found itself on the right side of several recent airline mergers. When Trans World Airlines, a major customer, bought out Ozark Airlines, a minor customer, business with TWA increased about 20 percent, Hoagland says. "I think our sales this year are up about 25 percent with most of our customers like TWA, Delta, Continental, and Northwest, who have bought other airlines," he says. "It's been luck more than design."

Forty percent of Racket's products are imported, mainly from the Far East, and that makes Hoagland a frequent flyer. "I'm constantly trying to find newly industrialized countries that can make some product cheaper than the existing manufacturers," he says, attributing part of Racket's success to his skill in sourcing. He will soon leave for Bangkok, Thailand, to scout for low-cost manufacturers; last year he traveled to mainland China on a similar errand.

Racket's rapid growth suggests a young company. However, it was founded in 1891 by Hoagland's great-grandfather, as a general store on the banks of the Missouri River. Hoagland's grandfather expanded Racket into the hotel and restaurant-supply business, and Hoagland's father made ties to aviation, serving as the hometown supplier to TWA beginning in 1947.

For a time, the company operated five retail hardware stores in Kansas City, but by the time Hoagland, 39, took over the company in 1980, the retail stores had been sold and the hotel and restaurant-supply business phased out. Today, Racket employs nine people, including one in a Japanese subsidiary, owns a fleet of trucks that transports products throughout the Midwest, and maintains warehouses in six cities.

36 Corporate Report/July 1987 PHOTOGRAPH BY MARK MCDONALD

Corporate Report 100, *July 1987, featuring Joe as he led rapid growth in Racket Merchandise Company. This was the second year in a row Racket was featured as one of the fastest growing companies in Kansas City.*

When Joe's father got into the airline business, everything he sold was made in America by companies like Harker China in East Liverpool, Ohio, and Jeanette Glass, made in Pennsylvania. Virgil sold glass salt and pepper shakers to numerous airlines in the United States and abroad. Joe remembers the metal shaker tops were made by Crowne Metal Company in Paterson, New Jersey, and the shaker bottoms were made elsewhere, and each item was shipped separately to Racket. He reminisces, "From the time I was eight or nine years old, I would spend summers and weekends screwing the chrome plated metal tops on small glass bottles. I asked my dad why he didn't sell these separately to the airlines, since they had to take the tops off anyway to fill them with salt or pepper. He said, 'That's not what they ordered.' And Joe said his dad never asked if they wanted to buy them separately," laughing and shaking his head.

"Made in America" was being challenged and changed, starting in the 1970s. As Joe worked with his father in that generational business transition, they both knew it was time to think, travel, and begin to buy globally.

CHAPTER FIVE

GOING GLOBAL—CATTLE AND AIRLINES

As Joe and I reminisce about the airline industry, and as we are talking about what is important to include in this book, I think about how much Joe's experiences in the global airline industry prepared him to take on the challenges of establishing a new American cattle breed. Communicating with people was a paramount skill we needed to explain our new breed and lay out the facts so that people would want to join us. We both had learned and used our communication skills, Joe in politics, and me in advertising. Joe was never intimidated and thought of resources in a bigger way than most of us do. He truly thinks globally. He isn't a person to just talk about doing something. He is a man of action, making things happen.

As Joe took over Racket Merchandise Company from his father, the products we were selling to airlines began to shift away from American made. Cheaper products came from Japan and Taiwan. Virgil's navy buddy from World War II had a son who worked for Wells Fargo Bank and lived in Taiwan. That son helped Virgil find new sources of supply there. But doing business on the other side of the planet had new and unique challenges, as well. It's interesting to think of the history of business communication from the 1970s to the present day.

Communication could only be through Western Union telegrams at that point in time. Telegrams! Phone calls wouldn't suffice for business details which needed to be in writing. So, telegrams were the first way to communicate that was quicker than the postal service.

Eventually, Racket got its own telex machine. This was a machine in which you typed your message on a paper tape with holes (no corrections allowed!) then fed this pre-coded tape into a machine that sent the message via telephone line to the recipient. Because we were charged by the minute for the time the message took to send, we tended to make them abbreviated messages. The telex tape would print out your message at the receiving end and also print a copy in our own office—not a Western Union office down the street. It was more convenient and seemed like great new technology at the time.

The next new technology we installed was a fax machine. That copied a page on the sending end of the communication, sent the message electronically, and printed out the document on the receiving end, on the other side of the world.

But emails and attachments became commonplace in the 1990s. Then came cellphones and video conferencing, which replaced personal sales meetings for much of our sales work.

Virgil took Joe with him to Asia early on, as Joe took over the family business and got out of politics. Again, we all went as a family on some of these trips to Taiwan and China. It was a challenge to travel with an eighteen-month-old on a seventeen-hour flight, but I enjoyed the experience so much that the tricky logistics were well worth it. (Newly marketed disposable diapers helped tremendously!) Family and business intertwined, as was our family legacy.

As time went on, manufacturing had morphed from one country to the next, as we always had to search for lower cost items to win contracts and supply products to the airlines. The chronology of our sources for our airline products

Virgil and Joe share a light moment at one of the many airports they were in together.

looks much like the trajectory of growth for other countries in the world over the last forty years.

From Japan and Taiwan, Joe sourced products in Hong Kong, Mainland China, Malasia, Singapore, the Philippines, and Indonesia. Traveling internationally had become a vital part of our family business. At the same time, that traveling and getting to know his suppliers gave Joe an understanding of different countries, different cultures, and he absorbed skills to get things done, in America and the world.

Here are some side stories that show what I mean.

As I said, the airline industry is global. Both our customers and our suppliers were worldwide. The passenger service products we sold were used in flight and included pillows, blankets, linens such as head rest covers, and stereo headsets for entertainment. These products were produced by light industry, often cottage industries, like cut and sew shops. Many tended to be hand work and labor intensive. Joe had to be creative to find new sources for these products constantly. This made for some interesting travel to rural areas in Taiwan, I recall, where small factories were set up in small villages, between houses and barnyards with ducks, geese, and the occasional pig. The Chinese are industrious.

Most airlines had some international routes and could take delivery of products outside the normal US customs borders. It became advantageous for Racket to do business in low-cost, under-developed countries but then deliver the products to the airlines' offices in hub cities, such as Singapore, Hong Kong, or Tokyo.

In Europe, in the 1980s, Joe found product sources behind the Iron Curtain where communism allowed factories to operate without true cost accounting, as long as they employed workers. Racket bought products from Poland, Czechoslovakia, and Hungary at a time when few Americans even traveled there. Those countries had centuries of glass and chinaware production with excellent quality and low prices. These Warsaw Pact countries welcomed us and our trade. We also sold items to airlines there, like LOT in Poland and Malev in Hungary.

In Hungary, we were such an important trading partner along with another American company, Coca Cola (!) that Joe was invited to meet the Prime Minister Karoly Grosz.

History has not given Karoly Grosz, Hungary's Prime Minister in 1989, the credit he deserves for helping to end the Cold War. He gave the order allowing East Germans to cross the Hungarian border with Austria, which then allowed them to travel on to West Germany. Once those East Germans began to seep into West Germany and tell their stories back home, it was only a matter of time until the Wall had to come down.

Joe has urged Hungarians to put a statue of Karoly Grosz in Budapest's

Joe shaking hands with Karoly Grosz, prime minister of Hungary, 1989.

Heroes Square, but Grosz is just another forgotten communist from an era they would rather forget.

Joe found an excellent chinaware factory in Hollohaza, Hungary, that became our supplier for TWA and American Airlines. Hollohaza made exquisite porcelain products, hand painted figurines, and demitasse sets. It was run by Laszlo Gomze, an academic who was a technical expert in producing porcelain. He was a communist, appointed by the Party to run this chinaware factory. We did business there for a number of years and have stayed in touch with him and his family.

After the collapse of Communism in Hungary, Dr. Gomze took a professorship at Miskolc University, specializing in nanotechnology. Meeting up with Dr. Gomze years later, he told us that when the Wall came down, he tried to buy the Hollohaza factory to keep it running after the Communist Party leaders had abandoned it. He was turned down by the new, local politicians who had questioned him and thought he was suspicious for doing business with an American company—us. "They told me they thought I was CIA," he said. Joe replied, "I always thought you were KGB!" And both men laughed.

On one early trip to Hungary to visit this chinaware factory, we took our two older children, Anne and Dirck, with us. We traveled by train, and when we reached the Hungarian border, the train stopped. Soldiers with guns came

Professor Laszlo Gomze, PhD and Joe, when we met again in Budapest, Hungary, in 2019.

onto the train and inspected us, our passports and our luggage. You can imagine this made a lasting impression on our astonished children. We stayed in a small, communist-run hotel and had to relinquish our passports so we could not leave without authorization. We were closely watched—at the factory and at the small hotel.

During this visit in Hungary, I went to the grocery store with the wife of the factory manager, just two mothers out to do some shopping. It was the first time that I had ever seen stores with virtually nothing on the shelves. How could a mother feed her family with nothing in the grocery stores? But there in Hungary, at the factory, our hosts were accustomed to scarcity. We were welcomed wholeheartedly and fed as well as possible and treated with great friendliness and respect. We treasure our memories of that visit. But I think of those empty shelves whenever I read about a country with communist or authoritarian rule. Politics matter and affect everyday life. Joe and I learned that early in our political life, and we continued to see the truth of that as we traveled in countries with different political systems. We don't understand people who don't want to think about or be informed about political issues.

Joe had traveled to Russia in 1985, when it was still the Soviet Union. Even in Moscow, store shelves were empty. He says the only green vegetable he saw there was a green pepper. The country was poor at that time. In 1989 when the Berlin Wall fell, Joe was driving to the chinaware factory in Hollohaza in eastern Hungary when he saw Russian soldiers who had dropped their rifles at the roadside and were hitchhiking back to Russia.

Oddly, much later, in 2016, we were invited to a New Year's Eve party at the Hermitage in St. Petersburg, Russia. There were many Russian oligarchs present. Putin was rumored to be expected but never appeared. By then, Russia was again a rich country, fueled by their oil industry. The streets were lined with upscale shops and luxury car dealerships. No more empty shelves. Brooks Brothers and Louis Vuitton goods were in the Gum Department Store, within sight of Red Square in Moscow. Many young Russians spoke in English to us.

Joe and his Korean supplier, Jae Rhee, in 2022. Together they sold more stainless-steel cutlery than anyone in the world in the 1990s.

The toast at the New Year's party was "May next year be as good as this year!"

Our stainless-steel cutlery supplier in Korea, Jae Rhee, also became a lifelong friend to Joe. He and Jae would consult each other about the state of the world and have long political talks. Joe has always joked that if he were president of the United States, he would appoint Jae as his secretary of state! However, neither was going to happen. But in the 1990s, Joe and Jae together sold more stainless-steel flatware than anyone in the world.

Throughout the late 1980s and early 1990s, Joe grew the company, traveling three out of every four weeks to meet new airline customers. I believe he has traveled around the world fifteen times. That doesn't count single flights to and from a specific place in Europe or Asia and then back. Delta airlines offered a special domestic ticket for twenty-one days of unlimited travel in the US. For a low fixed price, if you didn't go to the same city twice, you could travel for twenty-one days, to as many places as you liked. So, Joe would set out on these three-week sales trips, sometimes flying to St. Louis or Wichita and renting a car to drive home to us near Kansas City for a weekend. Then he would return to that city to resume the onward air travel on these special fares.

As we look back on those very busy years, it was a challenge for me at home, with three small children and the family ranches. I'll paraphrase Eleanor Roosevelt's line that "a woman is like a tea bag—you don't know how strong she is until she's in hot water." There were times that tested me. But I certainly

Norma and Joe in Maxim's restaurant, in Paris, France.

had a few "perks." With both wonderful parents and parents-in-law, I had a strong support system. I was able to leave the children with them for a few days at a time and occasionally meet Joe at the end of one of his round-the-world business trips for a few days to relax. The south of France was one of our favorite destinations to meet, on the warm and beautiful coast. Then we'd return to Kansas together. I am so grateful for the life I've had and these many opportunities to travel, meet people, and expand my knowledge of the world.

Prior to the terrorist attacks of September 11, 2001, there was no airline security. You would go to the airport gate just in time to board the plane, carrying a paper flight coupon for a specific flight, even if it was not in your own name, and still be able to board the aircraft. There were times Joe would buy tickets in one name and change who would accompany him on a sales trip later but be able to use the same ticket. Hard to imagine in these days of airline security!

All airlines offered meals in flight. Even for flights as short as Dallas to Kansas City, passengers were served a hot meal on a chinaware entrée plate. In fact, those oval casserole dishes were one of our main products. There was no food offered in the airports themselves back then. No need for meals because with no security, you could arrive and just hop on your flight immediately.

All that changed after the September 11, 2001, terrorist attacks. There were long security lines and earlier check-in times, leaving passengers with idle time in the terminals. Now there is plenty of time to eat before boarding flights, and no food is served on a flight that is less than four hours long. This had a profound impact on our airline business. We needed to look outside the USA for more customers for our products.

Racket opened offices in Hong Kong and Brussels to expand our sales worldwide. One of Joe's cousins from New York was John Vanderbeek. John was one of the original investors in Microsoft. We suggested Microsoft open an office in Hong Kong, to join us on the twenty-second floor, with all the other American companies, in the Lippo Center in the Admiralty District. Microsoft was across the hall from Racket Merchandise Company. Microsoft soon needed a bigger space, but Racket kept this same office for about thirty-five years!

While all this was happening, we were working hard to become successful cattle breeders. Although we had a series of ranch managers and cowboys, we both were very hands-on. We worked the cattle ourselves. We knew our animals' performance and their genetics better than anyone. We were learning how to breed for the traits we wanted. We were active in the Hereford groups in Kansas and nationally, attending the National Western in Denver. We owned the bull that sired two different back-to-back National Champion Hereford bulls in the early 1980s.

Appreciating Joe's political and legal skills, the American Hereford Associ-

RB L1 901 Domino 970, sire of two back-to-back Grand Champions at Denver's National Western Stock Show: L1 Pacesetter in 1983 and 1984, L1 Challenger in 1985.

ation named him a director of International Activities for the breed in 1985. This made sense, as Joe was traveling for business a great deal in these years of growth. While traveling overseas, he could represent the Hereford Association at cattle events worldwide, where he already had other business meetings. Thinking globally and having these worldwide contacts would certainly help us later, when it was time to promote Black Herefords.

We had consolidated our farmland and built our headquarters in Leavenworth County, with two and a half miles of frontage along the Missouri River. We farmed the rich bottomland along the river, growing crops to feed our growing cowherd. This ranch is an amazing place, with such diversity. We have pastures with improved grasses, brome and orchard grass, for the cowherd to graze. There are forested areas where we can log the native black walnut trees and hunt for morel mushrooms. The fields along the Missouri River are flat and fertile, perfect for corn silage for our cows and soybeans as a cash crop.

One of our favorite things to do, on one of those unnaturally warm days that happen in Kansas each February, is to drive down to the riverbank and have a picnic. We have nests of bald eagles along the river—amazing to watch. The margins along the river are so wild and untamed. We see wild turkey and white-tailed deer. Just sitting there and watching the river flow by makes us sense the ancient history of North America. We realize how incredibly fortunate we are to feel this tie to the land and our responsibility to be good stewards of it.

The Leavenworth ranch is listed in the National Registry of Historic Places as the archaeological site of the "Lansing Man." In 1902, the remains of an adult male and a child were found beneath twenty feet of undisturbed geological strata, which predates the last ice age. They are the oldest human remains ever found in the Great Plains of America. This place we call home has a legacy to protect.

High up on the bluffs of the Missouri River on our Leavenworth ranch is another tie to the past. The Concannon Cemetery is there, dating from 1855. Preserving it is another tie to the past for us. Standing on that land and looking to the southeast down the Missouri River Valley, we can see the skyline of Kansas City. It's a gorgeous view, but a stark reminder of how near the city is and how precious our agricultural footprint is.

At this Leavenworth headquarters, we built a sale barn facility, looking forward to holding our own production sales of our traditional Hereford beef cattle.

As our animal genetic selections for herd improvement became more sophisticated, it became clear we needed to utilize artificial insemination to expand the genetic base of our herd. So, while my mother watched our children, I went to artificial insemination school at Kansas State University's Kansas Artificial Breeding Service Unit. It was rigorous but only a two-day training session, with lots of emphasis on animal health and nutrition as well as training on the basics of insemination. I was one of only two women in the class of about twenty, and we both endured scorn from our instructor, unhappy that there were any females in his class. Of course, this made us both even more determined to learn. We both got our certification. Back home, I was happy to contribute this skill to help improve our cowherd's genetics and did our artificial insemination work myself for many years.

I needed help managing the cattle in Joe's increasing frequent and long absences, so our first resident hired hand was Kenny "Buck" Buchanan, a local Leavenworth legend. Prior to Buck, I had a neighborhood boy, Daniel Yunghans, help me after school and on weekends. Daniel's older brothers Zane and Bill were my emergency contacts if I had a crisis and was without Joe. One crisis I clearly remember was a huge ice storm that enveloped Leavenworth while Joe was gone. (Mrs. Murphy's law? Anything that can go wrong, will go wrong—when he's out of town!) I needed to bring my cows into the barn and feed them, but all the gates from pastures to barn were frozen solid to the ground. I couldn't budge them! Bill Yunghans came to my rescue, with a sledgehammer to break the ice, and then I could get my cows in to feed, open water, and shelter.

It was a challenge to be solo at times to get chores done with small children,

J&N Ranch Sale Barn, built in 1985, with ranch office and apartment built in front.

Feeding in winter is easier now with our "Big Bertha" Roto-Mix feed truck.

as any farm wife can tell you. I put a playpen in the barn to keep little Emma, especially, safe while I would feed or move the cattle. I made a little sign to hang on the inside door of the house if I was outside working while the children napped. Little Dirck and Emma couldn't read yet, so it was a picture of mommy and a cow. That way they would remember where Mom was when they woke up. Our children all joke they were indeed "born in a barn!" Joe was doing what he needed to do to grow our businesses, and I was doing what I needed to do to keep things running back at the ranch. Partnership.

Having consolidated all our cattle operations in Leavenworth, we held our first production sale in 1986, in our new sale barn.

In the 1980s we were heavily involved in showing cattle in the top Hereford shows like the American Royal in Kansas City, the National Western in Denver, and the Golden Nugget in Reno. Often, we would consign our entries with professional fitters to improve the odds of success. We produced Champion, Reserve Champion and numerous class winners wherever we went with our traditional Herefords.

At the time, we did notice that the two premier Hereford breeders, Les

First annual production sale at J&N Ranch, 1986. Joe is on the left, Stanley Stout, our infamous auctioneer in the center, Craig Ludwig of AHA on the right, clerking the sale. Craig would later come to work for us at the ABHA office.

Holden and Jack Cooper, both of Montana, never showed their cattle. The offspring of their genetics had plenty of showring success. They left that part of the business to their customers. They concentrated on their own genetics, their own ranches, and their own production sales to make money. We realized that was how real impact was made in the beef genetics business, by paying attention to your cattle and forgetting about the showring. We remembered this clearly, and that was our approach when we began to produce Black Herefords.

By 1989 our traditional Hereford cattle operation had peaked. We produced a bull that was Reserve Grand Champion at the American Royal as a yearling. We took him to the National Western Livestock Show in Denver a few months later. We sold the bull at the sale there for $5,000 which seemed like a huge amount at the time. We stayed at the historic Brown Palace Hotel and felt we had arrived in the cattle business.

But things never stay the same. As market forces changed, we began to consider a change. The Angus Association had begun its "Certified Angus Beef"

Joe and Norma with our yearling bull champion at the National Western Stock show, Denver, 1989, pictured with Bob Brink, professional show fitter.

promotions, a brilliant idea that placed a premium on cattle with black hides, emphasizing the meat quality of the Angus breed.

We began to see that our traditional Herefords weren't easy to sell any longer. We had begun our herd with excellent Hereford Line One genetics from Miles City, Montana. Even those good-performing cattle were not enough to overcome the financial downside caused by the Angus market premiums. Red-hided cattle began to be discounted at the markets. Buyers wanted black-hided calves. Time for an adjustment. We decided to also raise Black Angus cattle.

In 1992, Richard Spader was the CEO of the American Angus Association. He signed our membership certificate. Perhaps no one until Wade Shaffer, CEO of the Simmental Association, had a greater impact on beef production in America. Richard Spader was instrumental in starting the forward thinking, branded beef program known as Certified Angus Beef. It changed the marketing rules for cattle in America, providing premiums for cattle with black hides and discounts for red hides. His son, Brett Spader, had worked ringside at our production sales and later became CEO of the American Charolais Association. Brett was instrumental in helping to start CattleTracs, another Hoagland business which I'll talk about later.

Once we started a registered Angus cowherd on our ranch, it added a new set of forms and deadlines, and a new system to learn. It also brought a change in our handling techniques. Unlike the docile Herefords, with Angus cows there was a more aggressive temperament. We had to be more careful with the children helping with chores and working the cowherd. No more playing in the pastures with the cows!

It was always a challenge to get to the top as a registered breeder of seedstock for the commercial cattle industry. We essentially were starting over with Angus cattle. You had to do something to "leapfrog" ahead of other breeders, Joe knew, but we hadn't quite figured out what that was yet.

It was time for us to start something new—a new cattle breed, as it turned out, looking back. But, before that, there was one more career move for Joe.

CHAPTER SIX

THE BASEBALL DIVERSION

Under the capable management of John Postlewait, Racket Merchandise Company's number two, the company continued its string of successes. John had worked for Joe's father since high school. After graduating from Creighton University and a stint in the army, John returned to join the management of our family business. He was already in place when Joe returned from the state legislature and his law practice to run Racket, starting in the 1980s.

The next several years were ones of rapid growth for Racket. Skillful hiring and putting people in the right places to grow the business, Joe began to have more free time of his own and didn't have to travel internationally as much. Spending more time with us at the ranch and building the cow herd took some of his time, but Joe was accustomed to doing several things at once. He considered starting another career in professional baseball, following a dream and hoping to become a general manager in the Big Leagues.

With a law degree, which would be a plus as a GM, and with political and business management skills, Joe had many of the qualifications needed. He just needed to develop player evaluation skills, so he became a "bird dog" scout (an unpaid position) for the Houston Astros under the tutelage of Bill Kelso.

Joe had played baseball all his life, starting with that hand-made diamond his mother had built for him and his friends. He had gone to the Mickey Owen Baseball camp in southwest Missouri numerous summers as a young boy, returning as a counselor when he got older.

Joe was enamored of baseball scouts ever since he met Tom Greenwade of Willard, Missouri, while at Mickey Owen. Tom was perhaps the greatest baseball scout of all time. He recommended Jackie Robinson to Branch Rickey when he worked for the Brooklyn Dodgers. After Greenwade was hired away by the New York Yankees, he signed Mickey Mantle, and many other Yankee stars in the 1950s and 1960s.

Ad for Mickey Owen Baseball School, Miller, Missouri. Joe and our son Dirck both attended camp there for numerous summers.

So, Joe knew this game of baseball inside and out and loved it. Learning the art of scouting from Bill Kelso was really fun for Joe. Bill, Joe's scouting tutor, grew up in Kansas City and played four seasons in the Major Leagues. He and Joe played ball together in a local semi-pro league after Bill had left the Cincinnati Reds. By the time the scouting lessons started, Kelso was a scout for the Astros and later became their scouting director.

Joe learned how to evaluate players, using the ranking system the scouts used. He got a radar speed gun to measure the speed of pitches. This was a whole new level of baseball, and Joe was in his element! I was happy he did something for himself, as he worked so intensely for the family, for Racket, for J&N Ranch.

After learning player evaluation skills from Bill, Joe got a part-time job (paid by the game) with the New York Yankees. The next year, the New York Mets hired Joe full-time, with salary, to travel several states, looking for talent and to attend spring training with the team.

Joe stayed active in the family business but was only needed in the mornings a few days a week to sign checks and confer with John Postlewait. Afternoons were free to go watch baseball games. He traveled over the Midwest region but usually within a day's drive.

There was a highly touted high school center fielder playing for Lee's Summit High in the spring of 1996 when Joe went to see him play. Lee's Summit was playing Fort Osage High School, and Joe spotted their sixteen-year-old third baseman. His name was Albert Pujols. Joe was the first scout to see Albert play and continued to follow him from his years at Fort Osage High through his one year at Maple Woods Community College. This would have been for the 1999 Major League Draft.

Joe sought approval from the Mets to offer Pujols $100,000 as a fifth-round pick. Albert agreed he would sign. This was part of being a scout. Teams did not want to use a draft pick on someone who would not sign with the team. So, you had to get an agreement to sign for an amount both sides would accept.

When the fifth round came, the Mets skipped and continued to pass over Albert until the St. Louis Cardinals took him in the thirteenth round and paid him $60,000. His first year in the Minor Leagues, Pujols was that league's Most Valuable Player. During spring training the next year in Florida, he made the major league team and finished the year for St. Louis as the National League's Rookie of the Year in 2001.

Albert played twenty-two seasons in Major League Baseball. He became one of the greatest power hitters of all time, driving in more career RBIs than Babe Ruth. Albert sits at number two all-time in RBIs and is only the fourth player in MLB history to hit over seven hundred career home runs.

The next year, the Mets took Joe's recommendations more seriously.

The Mets used their first-round pick in the 2000 draft to take a high school pitcher from St. Louis that Joe had first seen as a fourteen-year-old. His name was Bobby Keppel from DeSmet High School. The Mets paid him $895,000 to sign. He reached the Major Leagues playing for several teams and finished his career in Japan.

Our son, Dirck, also had a love of baseball—is it genetic? He attended Mickey Owen Baseball School, as well. When Joe was still playing baseball at the age of forty in the Men's Senior League, Dirck was big for his age. He was fourteen, six foot three and 165 pounds. Joe's team would play in tournaments in Chicago, Orlando, and Phoenix, to name a few. The whole family would follow along on road trips. Since Dirck looked older, Joe's teams would have Dirck pitch. He pitched against the Lackland Air Force Base team and several small and junior college exhibition games. It was great preparation for Dirck. And that year, Joe's team went on to win a National Championship in the Senior League.

Joe and Bobby Keppel, the New York Met's first-round draft pick in 2000.

Not only did Dirck pitch in high school, but he was invited to tournaments and games where the nation's top baseball prospects are invited to compete against each other. These showcase events like the Area Code Games and Perfect Game tournaments were watched by the colleges for scholarships and scouts for Major Leage Baseball. It also gave players in a region a chance to get to know each other.

As Dirck was playing his last summer before going to college, it was inevitable that he would pitch against Albert Pujols who lived nearby in Missouri. He faced Albert three times that summer. Albert never got a ball out of the infield and Dirck struck him out twice. That is my favorite family legend, as a proud baseball mom. Another is while in a high school game, he set a record that may still stand in all of baseball. He struck out five batters in one inning. Not only could the batters not hit his pitches, but his catcher couldn't catch them, either! Two batters got on first after dropped third strikes.

Dirck played at Tulane University, which had the best record in Division I college baseball in 2001. Dirck was the winning pitcher when Tulane set the school's winning streak record. He shook hands with President George W. Bush, himself an enthusiastic baseball fan. President Bush threw out the first

At center right, you can see Dirck's back (HOAGLAND) as he shakes hands with US President George W. Bush. The president threw out the first pitch for the start of the College World Series, Tulane vs. Stanford, 2001.

pitch to begin the College World Series in 2001. Today Dirck is recognized in the Tulane Hall of Fame, where he played from 2000 to 2003 and was part of that College World Series team in 2001.

Then, on September 11, 2001, the terrorist attacks happened in America and Joe's baseball career came to an abrupt end.

Airlines were ordered to immediately remove all stainless-steel cutlery and replace it with plastic on all flights. We were faced with a warehouse full of products that the airlines suddenly did not want to buy. Although Racket had never missed a loan payment, several banks that provided lines of credit became nervous, and Joe had to calm their fears as he worked day and night to avoid bankruptcy.

While this was happening, we noticed something interesting was happening in our pastures. The registered Black Herefords that we had purchased just a few years earlier were outperforming everything else, both the Angus and the traditional Herefords. It was time to start a new cattle breed.

CHAPTER SEVEN

WHERE IS THAT SHOEBOX?!

Where do you begin to start a cattle breed? You'll need a breeding protocol, a system to register individual animals, and you'll need to gather performance data on each animal. You'll need to organize an association for the members and hold meetings for the members to gather and share experiences. Perhaps you'd even need a magazine and especially a website to communicate this new breed to the world. Time to be brave; time to get busy.

Joe and I dared to believe that we had an idea to improve America's cowherd by developing the Black Hereford beef cattle breed, combining Herefords with a small percentage of Angus genetics. We believed, and still believe, it would be the perfect breed to use on America's largely Black Angus cowherd. Using a Black Hereford bull would give ranchers a reliably black white-faced calf to capture those higher prices on the market given for black-hided cattle. Because of the effective "Certified Angus Beef" marketing, red-hided cattle were being discounted at the markets. So here was a way to get a reliably black-hided calf on a commercial, mainly Angus cowherd. Many of those commercial Angus cows still carried a red hair gene and still had red calves. A Black Hereford bull would eliminate those red calves.

Black Herefords would also add exceptional feed efficiency and docility, known and proven hallmarks of traditional Herefords. These two traits go hand-in-hand, as animal scientists have proven. A calm disposition leads to better feed efficiency. Not as much energy goes into maintaining a calm animal as a flighty one. More beef from less feed. Plus, they are simply easier to handle.

Black Herefords have remarkable adaptability to different geographic locations, again from their feed efficiency. They thrive at high elevations and low, from harsh winter conditions to hotter summer climates.

Black Herefords would also give the boost of heterosis to America's mainly Black Angus cowherd. Being predominately composed of Hereford genetics, that genetic boost you get from combining two different gene pools or breeds is a huge offering to commercial cattle ranchers. It was the answer, we believed, and still believe. It is the perfect cross for the American rancher and America's largely Angus cowherd.

But that answer began with our friend John Gage. John's father had been mayor of Kansas City in 1940, and an attorney, founding the law firm of Gage and Tucker. John B. Gage, the father, had always been involved in agriculture. His favored breed was milking shorthorns.

Our friend, John C. Gage, the son, was a big, powerful man with a gentle demeanor, a graduate of West Point. He followed in his father's footsteps, became an attorney, and practiced law with Jim Pearson who would later become a US senator from Kansas. John represented the school board when Joe's father was its president.

John owned Blue Jacket Hereford Ranch near Eudora, between Kansas City and Lawrence, where he lived with his wife, Eleanor. They, along with their son David and his wife, Mary, raised Polled Herefords and ran a bull feed test program.

In his law practice, John Gage handled primarily cattle and ag issues. He had definite ideas about cattle breeding and genetics. He didn't approve of show cattle, and it didn't take long to find this out in a conversation with him. That is interesting, given the fact that he was at one time the president of the American Royal, the premier national cattle show hosted in Kansas City.

In his capacity as an attorney, John C. Gage represented the American Polled Hereford Association (APHA) during merger talks with the American Hereford Association (AHA.) The AHA had recorded the British breed pedigrees of Horned Herefords since its founding in America in 1881. It had a long history and tradition, slow to change with the times.

Other breeds had begun to turn their cattle's hair black by crossbreeding with Angus to capture those "Certified Angus Beef" market premiums. Limousin, Gelbvieh, Simmental, Saler, even the all-white Charolais breed had talked about this transition, according to our friend Dr. Bill Abel, their CEO at the time.

As the AHA and APHA merger negotiations were underway, John Gage presented the idea of turning Herefords black, to accept them as a sub-registry in this new combined association. He used that as a negotiating tool. It's not clear that he truly intended to start up this new breed. John was able to successfully navigate these meetings and convinced the boards to consolidate the two Hereford breeds into one association. They would be financially stronger together, rather than separate, but the new association was formed without the third arm, Black Herefords.

The American Hereford Association and the American Polled Hereford

Association formally merged the two breeds into one association in 1995. It was to be known as the American Hereford Association.

While representing the Polled Association as their attorney during the merger talks, John Gage announced the formation of the American Black Hereford Association in a letter dated February 5, 1994. (See Appendix A) This letter was sent to the AHA Headquarters, to H. H. "Hop" Dickenson, executive vice president of the American Hereford Association:

> Dear Hop:
>
> I enclose herewith a copy of the Articles of the above (American Black Hereford Association,) which have been recorded in Kansas. The name has been reserved in Missouri. No rules have been drawn up.
>
> We have no intention of promoting such an association. Its purpose is primarily defensive and for the time being at least it will be private and not open to all breeders. It will be used to develop a few black whiteface cattle of what we consider the best available breeding to compete with the better registered Angus bulls that at this time are winning the competitive battle for bull sales in our area.
>
> I do not have the confidence you and some others have in changes that will result from the black cattle being offered by continental breeds. It could have the effect of pushing a lot more commercial breeders toward straight black Angus operations unless they can see a clear alternative, as has been indicated by conversations I have had with several successful commercial producers.

John Gage never raised a Black Hereford.

While that idea of a Black Hereford sub-registry was rejected during the merger talks, John had set up a Kansas corporation to legally organize a Black Hereford association, as described in this letter to Dickenson. He announced it as being held privately by himself and not as a member-owned association, like AHA and APHA.

When Joe and I purchased the corporation from John in 1997, he wrote to us the following (See Appendix B) on November 6, 1997:

> Dear Joe:
>
> As you will note we did not adopt bylaws because they have to be tied into the rules of membership and registry. I include in the enclosed some reference material you may want to look at in developing such rules. It would probably be advisable to get rules from other associations like Brangus, Gelbvieh etc. to see how they have handled setting up a registry for crossbreds. My tentative idea about it was to record any Angus-Hereford crosses with registered ancestry, then have them move to a full registry when they were perhaps at least 5/8ths Hereford, with a special registry for those in full registry that are proven by DNA testing or otherwise (a lot of progeny) to be homozygous black. Some performance requirements should also be involved, maybe reporting weights on all calves.

As John Gage suggested, when we wrote the first bylaws for ABHA, we followed the breeding rules of the continental breeds that had turned their hair color black. Everyone used the five-eighths rule as suggested by the Beef Improvement Federation (BIF). They all started with an F1 female cross, which was 50 percent Angus and 50 percent Simmental, for example. F1 bulls were not suggested in the next steps in the breed-up process. The Angus-Hereford crosses with registered ancestry would move to full registry when they were at least five-eighths Hereford, again as John Gage suggested.

Our intention was always that this should be an association owned by the members, as all the major breed associations were. We would not keep it private as John Gage had done but make it accessible to fellow breeders to run. But we needed time to organize the association and get the rules set, so taking it public would be in the future.

Along with the corporation, we purchased from John Gage two F1 females bred to his Hereford bull 51M. One produced a bull on our ranch known as BJH Balder 7504 and we recorded it as the first registered Black Hereford in the association in 1997. Even though this bull was born on our J&N ranch in Leavenworth, Kansas, we gave the honor of this first bull to John by giving it the prefix BJH for Blue Jacket Herefords, John's ranch. "JN Balder" for bulls or "JN Baldee" for females were the names we decided on for all the animals born at J&N from then on. The other cow purchased from John gave birth to a heifer calf, and she was named she was named BJH Baldee 7901, again giving the honor to John Gage.

John's fellow Polled Hereford breeder, Frank Felton of Maryville, Missouri,

Rule 2.2 Eligibility for Registration
Black Hereford animals of any age for which application for registration is made may be registered with the Association upon proof of compliance with the following requirements:

A. The animal must be a Black Hereford, as determined under the Rules of the Association.

1. Therefore a bull or cow must EITHER:
a.) Have an ABHA registered sire and an American Hereford Association registered dam.
OR
b.) Have both ABHA registered sire and dam.
OR
c.) Be the product of a pairing described in **a.)** that was not able to be registered due to hair color and an ABHA registered sire.
OR
d.) Be an American Hereford Association sire approved for admission by the ABHA executive officer.
OR
e.) Be the product of an ABHA sire and a F1 dam whose parentage is registered in the American Angus Association and the American Hereford Association.

Rule 2.2 of the American Black Hereford Association bylaws. This breeding protocol was recommended by the Beef Improvement Federation and suggested by John Gage when we bought the association papers from him.

Oil painting of the first registered Black Hereford which was born on J&N Ranch, painted by longtime time Hereford artist Bud Snidow.

had joined him in the idea to breed-up a Hereford with black hair coloring. By breed-up, I mean you followed the breeding protocol as described earlier from BIF. From an initial cross of registered Hereford and registered Angus, you got an animal half of each breed, or an F1. Then you bred that F1 back to a registered Hereford, resulting in a three-fourths Hereford calf. Take that three-fourths blood black-hided animal and breed it to a black-hided half blood (or F1) and you produce a black-hided five-eighths blood Black Hereford which could be registered. Maintaining the black hair color while adding an additional percentage of Hereford genetics is the goal. This method is how all the other beef breeds had turned their hair color black, all while concentrating the original breed's genetic characteristics.

Frank Felton was a real character, laser-focused on his cattle breeding. We visited Frank at his ranch and it's a vivid memory for me. Frank simultaneously chain-smoked, chewed tobacco, and talked-nonstop. He showed us his cattle and gave statistics on each from memory. His office was lined with photos of cows and bulls in his herd, all behind nicotine-yellowed glass. The term "fanatical" comes to mind here. He had that hint of genius, and he knew his cattle! They were fertile, thick, and strong. We bought two bulls from Frank to lay the foundation for our own Black Hereford herd.

John Gage's health failed further, and he passed away in 2000. Interestingly, Frank Felton also died not long after, in 2003. It marked the loss of two great, forward-thinking cattlemen.

Anyway, in 1997, Joe and I had purchased the papers and records from John Gage for the ABHA and two cows, bred from his registered Hereford and Angus stock. Believe it or not, the association papers and hand-written pedigrees were given to us in a shoebox!

We planned at that time to just put these two cows into our commercial herd. The idea of truly starting a new breed had not hit us yet. Things were busy with our three children, all in 4-H, sports, and with all their school activities. Joe was still traveling a lot for business—thinking back, how did we manage it all?

In addition to the Kansas ranches, we ran eight hundred commercial cows on a ranch in Montana, run by our friend and partner Gary Bishop. Gary had been with the Hereford Association as the director of youth activities, in the 1970s, and we had become friends with him and Vickie, his wife. Gary had left

Our children Anne and Dirck, showing cattle in 4-H.

AHA in 1983 to manage their family ranch in Montana. J&N supplied the traditional Hereford bulls to use on their cows and brought our share of the heifers back to our ranch near Eskridge, Kansas, each year. We sold these commercial bred heifers in our annual production sales.

One of our favorite family vacations during those years was going to Montana for the gather and branding at Brooks Ranch. Our children and the Bishop children got to be friends. We would help gather the mother cows and their babies, do vaccinations, and branding. It was done in the old Western style, all on horseback, with calves roped and brought to the crew to be worked. Our children have great memories of these trips, one of which is Grandma Janet Brooks's famous cinnamon rolls and cookies, brought out to the pastures to feed hungry young ones!

As we brought these heifers back to Kansas, we would keep some in our commercial herd. We watched them grow, become cows, and as we evaluated their calves, we realized something. The Black Hereford registered cows originally from John Gage and their offspring were truly superior. Having registered stock on both sides of the pedigree really mattered.

Joe said to me, "Where is that shoebox?!"

In the words of David McClellan, our feedlot consultant from Fremont, Nebraska, "Take advantage of what we know about Hereford heterosis without indiscriminately using Hereford genetics. Look for Herefords that can bring something else to the table. Black Herefords can do that." Joe and I worked hard to research and find the best genetics to lay the foundation for the breed.

We began to see the future: better market value, better Hereford. Joe has always been a forward thinker. Some of our fellow breeders called him "visionary." I can tell you, he thinks in his sleep. I never know what new idea or business will be presented to me over coffee each morning. I pedal fast just to keep up. But Joe has always encouraged me, believing I could do things throughout our life together that I had never thought I could do. Isn't that the perfect partner?

Having a registered Angus herd, a registered Hereford herd, and now the start of a Black Hereford herd, the recordkeeping became daunting for me. Each association had its own separate forms, deadlines, and requirements to register calves when they were born. I learned to be organized! As I kept the data for each set of cows, we saw the outstanding performance of these first Black Hereford cows. We began to add more registered Hereford cows to our Black Hereford herd, carefully matching the genetics to strengthen their desirable traits.

As I mentioned before, the original pedigree papers and printed association legal documents had come to us in a shoebox. John Gage's pedigrees were hand-written. For the time being, I did the same, recording all registration numbers for the sires and dams and keeping all the dates, names, weights, and data recorded by hand. Soon we knew we had to find a computer program to make this process easier, more accurate, and faster.

We found there weren't any affordable computer programs available for beef cattle registration. The program used by the Hereford Association cost tens of thousands and was based in Australia. We weren't ready for the "big leagues" yet!

First, we tried Redwing Software, which was designed for dog pedigrees, I seem to remember. It had such limitations and poor customer support that we discarded it quickly. We paid for everything personally, so this was a blow to us.

Centric Software became our first successful computer program. Working with its founder, Jerry Young, we utilized all of Centric's capabilities. Long weeks, days, and hours were required of me to input all the animals, past and present, into this new computer system. Working with the Hereford and Angus association papers, I entered all the proper registration numbers, names, birth dates, data, etc. for each animal, regardless of breed. At least three generations back needed to be recorded on each animal's pedigree. This was important to show each Black Hereford was from registered stock.

The association remained private as J&N Ranch used its registered Hereford and registered Angus cattle to start raising Black Herefords. By February of 2003, we were selling about twenty Black Hereford bulls in our annual bull sale.

Using information gained by his international business acquaintances, Joe knew what steps to take next. In April of 2003, we applied to the Food and Agriculture Organization of the United Nations (FAO), based in Rome, Italy, to accept Black Herefords as a separate breed of cattle. We were approved later that year.

It had been nearly fifty years since a new breed had started in the USA.

Then on March 20, 2003, the US invaded Iraq. Starting the Black Hereford Association would have to be put on hold. Joe was getting back into politics.

CHAPTER EIGHT

RUNNING FOR THE US SENATE

In the first election I was eligible to vote, I voted for George McGovern and opposed the war in Vietnam. Joe voted for Nixon and supported the war. Joe's family had always been Republican, as had mine, typical of Kansas. But I had seen too many of my high school classmates get a low draft lottery number and go to Vietnam. Many didn't return.

Kansas has always been a very Republican state. When Joe was majority leader in the early 1980s, he was a member of the Executive Committee of the State GOP. He had been a Speaker at the Republican State Convention. Bob Dole had named Joe along with John Peterson, another young Republican legislator, as the "future of the Republican Party in Kansas" during a speech at the State Convention in 1974.

By the Spring of 2003, Joe had been out of politics for nearly twenty years. He was back in the airline supply business full-time and doing business in the Middle East, among many places. George W. Bush was president of the United States.

One of Joe's biggest airline customers in the Middle East region was Saudia Airlines. In a meeting with the head of purchasing, who was also a member of the Saudi Royal family, the conversation turned to President Bush's build up for the invasion of Iraq. The Saudis were worried. They feared toppling Saddam Hussein, who was a member of the Sunni minority in Iraq, would tip Iraq's Shia majority toward Iran, a Shia stronghold. They worried the void created by the loss of Saddam Hussein's hold on power would destabilize the Middle East. Joe said not to worry, that Bush was smarter than that. "He is only doing it [building up troops] to get inspectors in to look for weapons of mass destruction," said Joe.

The invasion began on March 20, 2003.

Under Article 5 of the NATO Treaty, members agree that an attack on one is an attack on all members, and they will defend together any and all NATO countries. NATO joined the United States in the invasion of Afghanistan to hunt down Osama Bin Laden and his terrorist network after their attack on the USA on September 11, 2001. The case of invading Iraq was different.

Jacques Chirac, French president and fellow Tulane University graduate, blocked the USA from using Article 5 to make the invasion of Iraq a NATO action. Although Article 5 had been used against Afghanistan in October of 2001, following the USA terrorist attack, as I said, this was different. France had colonized North Africa in the past and knew the history and culture of the region. They knew Iraq was a Shia Muslim nation. Chirac said the Shia had nothing to do with 9-11. It was Sunni extremists from Saudi Arabia that had committed the terrorist attack. This was known.

No weapons of mass destruction were ever found in Iraq. These military actions were hard for us to watch. Joe knew the difference between Sunni and Shia in Islam and wondered why no one in Washington hadn't given better advice to Bush.

Joe has known every Kansas governor on a first name basis since John Anderson—twelve in all. Kathleen Sebelius was governor when the Iraq War started. She had previously been the lobbyist for the Kansas Trial Lawyers who supported Joe and the changes he had made to Tort Law in America. Joe asked Kathleen, a Democrat, if he could speak at the next Democrat State Convention to express his opposition to President Bush's actions in Iraq. Larry Gates was the Democrat state chairman and a classmate of Joe's in law school at KU and gave his approval.

Joe spoke to the Kansas Democrat Convention in 2004. He gave his speech to a standing ovation. He said the war in Iraq was unnecessary.

Kansas had not elected a Democrat to the US Senate since 1932 and has sent fewer Democrats to the Senate than any state in the nation. Following this speech, we got a phone call from Chuck Schumer, who at the time was chairman of the Democrat Senate Campaign Committee, asking Joe to run against incumbent Kansas Senator Sam Brownback. He had Alan Seacrest, a Washington pollster, do a poll in Kansas on Hoagland vs. Brownback for Senate. It would become a targeted race.

Joe and I were in Brussels, Belgium, when Schumer called again, inviting us to come to Washington to meet with Harry Reid, senate Democratic leader. So, we got on the next flight and went to Washington DC. We arrived in a snowstorm and stayed in the Hay Adams Hotel directly across from the White House. The next morning, we went to the Capitol for a fireside chat in Reid's office where a US Marine stood guard. It was a magnificent office with windows looking out over the majestic snowy landmarks. We were seated by a wood burning fireplace. Reid encouraged us to run. He said they would throw fundraisers for Joe in New York, Miami, and Los Angeles. Joe would have the money he needed for the campaign.

Joe said he wanted more from the party. He said, "Democrats need to talk more about the War in Iraq, the false intelligence, and the unnecessary casualties. Democrats need to tell the American people that Republicans lied to them." Reid said," Well—but we have Democrats who support the war. Some of our friends support the war." Joe asked, "What do you mean, friends?" Reid said, "You have to make the Iraq War an issue in your own campaign with the money we give you." Joe replied, "Oh, you mean the defense industry."

That is when Joe and I decided being in the US Senate would not make a lot of difference. We were reminded of the farewell address of a fellow Kansan, Dwight Eisenhower, who said, "Beware of the influence of the military industrial complex."

Bipartisan support for war efforts in Congress is directly correlated to the location of defense contractors and military facilities in congressional districts throughout the USA. Big money runs American politics and policy. That's what makes change nearly impossible.

So, politics was put aside as we turned our energies again to the Black Hereford Association. It had been privately owned since 1994. The breed had now been officially recognized as a separate cattle breed. It was time to start selling memberships and open the association to breeders other than J&N Ranch. We had seen what these cattle could do and wanted to share it with other seedstock producers and commercial ranchers.

If Joe had run for the US Senate and won, there would be no Black Hereford Association today. We had our time in politics, but this was not the time to return to it. Reflecting on that period of our lives, we know we have made the right choice. Joe said that starting the Black Hereford breed will be his most noteworthy achievement in life.

CHAPTER NINE

BLACK HEREFORD ASSOCIATION BEGINS

Back home in Kansas after that flurry of political and personal decision making, we planned how to begin turning the Black Hereford Association from a privately held company into an association that would be open to members, with cattle sales and events for the members. We sought advice from trusted, longtime friends.

Before selling memberships in our new ABHA, Joe had lunch with Gary Bishop, our friend from Montana. Gary, Vickie and their girls had moved back to Kansas City for Gary to work for the American Hereford Association in 1999 as their director of breed performance and field services. Joe asked Gary again to approach the AHA board, as John Gage had done, to approve Black Herefords as a sub-registry in their association. Joe thought it could be a future source of revenue for AHA and this might be a good time to combine the associations. The only Black Herefords at that time were on J&N Ranch, our Leavenworth home along the Missouri River. Gary tried to do this for us, but the American Hereford Association board wouldn't agree.

Gary Bishop passed away December 17, 2002, and we greatly missed his friendship and counsel. After a time of recovery, we asked Gary's wife Vickie to come to work for us to help start this new breed. The ABHA work was beginning to be more than I could manage with the ranch and three children in school and activities. It was also time to get the Black Hereford Association off my kitchen table, as the saying goes, and become more professional.

Vickie Bishop had grown up with Hereford cattle and knew how to manage registrations and performance data. She helped us sell our first memberships and organize our first sales. At a critical time, she recommended an auctioneer, Tommy Barnes, to us. Vickie was a great help at a pivotal point in the growth of our breed. Vickie cautioned us that we were attracting mainly "novelty and hobby breeders," not serious performance breeders. She was right.

Initially, new cattle breeds attract novelty breeders. It's a phase new animal breeds of all kinds must go through. These early members join the association to be the "first" in their area to raise something new and different. When the novelty wears off, they often disappear. Our initial members in the ABHA are nearly all out of business today. The challenge for any association is to survive long enough to attract professional breeders who raise cattle for their income. A cattle breed cannot survive in the competitive world of beef genetics unless it makes a real contribution to beef production. Genetic evaluations based on generations of performance recordkeeping and tracking DNA markers for certain traits have replaced the showring, where selection was based on the animal's appearance.

Our bulls and the breed are backed by science. Measuring tools that evaluate cattle by the numbers is what sells our cattle today. We had been breeding registered Hereford cattle since 1978, collecting performance data. Starting in 1992, we started also raising registered Angus cattle, measuring the same performance traits. So, when we started combining these two groups of cattle together to produce Black Herefords, we already had many years of performance and breeding records to base our genetic decisions on to create a new cattle breed.

Without that jump-start of data to start to quantify the value of Black Hereford genetics, the new breed would have stalled out after that first phase of development. We were thinking for the long term and wanted staying power in the competitive world of commercial cattle. Early on we needed to attract at least some professional cattle breeders.

As time went on, we added features to Centric Software to record more and more performance data. Jerry Young worked with us and the specialists we consulted as we added Expected Progeny Differences (EPDs) that gave us comparisons to evaluate animals. Expected Progeny Differences are a genetic evaluation tool developed by animal scientists to measure how likely a trait might be passed on to an animal's offspring. It's a bit like a guarantee of what you are buying.

These EPDs could work to compare animals within our herd or against another breeder's animals, once we had fellow Black Hereford breeders. We weren't yet able to compare our data with other breeds, but we felt sure that would come in the future, a future we were beginning to see for our fledgling breed.

Joe seriously began a push to advertise the concept of Black Herefords. From his political experience, he knew how to contact different press groups and pique their interest so they would write about this new concept. He then began to lobby to make Black Herefords a recognized, separate breed. We needed this for, among other things, labeling semen to sell in the USA and worldwide.

The first American Black Hereford ad.

In 2003, thanks to Joe's persistence, Black Herefords had received an international breed designation from the International Animal Breeds Association in Rome, Italy, under the auspices of the United Nations's Food and Agriculture Organization (FAO). Our designation is "HB" for Black Hereford. We were added to the designations of "HH" for Horned Hereford and "HP" for Polled Hereford, both previously recognized as separate breeds. This code would be printed on each straw of semen to ensure customers knew which breed they were buying and using.

Because we were thinking globally about our airline business, we naturally wanted to promote Black Herefords internationally. Joe knew how to reach out to the FAO in Rome to get that international breed designation. As a result, Black Herefords were immediately recognized worldwide and not just in the USA.

Also in 2003, we had decided it was time to hold our first Black Hereford sale at our ranch. We had been advertising for a while in some printed cattle magazines and newspapers. We also paid to have a website built for people to see the advantages of raising this new breed of cattle. The internet was a big help in spreading the word about Black Herefords. But—it was still a leap of faith to see if others would join us.

In our first advertising pamphlet, we pointed out the economic advantages of Black Herefords. We handed them out at livestock shows and cattle conventions. Black hides, darker eye pigment, udders, and scrotums with black pigmentation were all economic traits important to commercial cattle producers. Black hides brought market premiums and higher prices. Eye pigment led to less tendency to get eye diseases or susceptibility to infections from flies. Black udders and scrotums mattered in cold, snowy climates, where sun reflected off the snow led to sunburn on white udders and scrotums. All these were important financial traits that Black Herefords offered.

These traits were something new for Herefords and would add value to the cattle industry. Another positive genetic surprise was waiting for us as we developed Black Herefords, in the future.

But back to the beginning of the American Black Hereford Association . . .

In 2004, we opened the ABHA to other breeders. It had been ten years since the ABHA was incorporated and not in private hands.

At our production sale that year, we held a small meeting for prospective breeders in our barn office. About a dozen people attended and all joined at that initial meeting. We realized this would become a "team of rivals," a term coined by Doris Kearns Goodwin, one of my favorite authors. We all would be competing for customers eventually. But in these early days, we needed to work together to promote the breed. There was great excitement in this small group of breeders from several different states. Initially, there was great camaraderie.

Cover of the first American Black Hereford Association brochure.

PLUS *Color You Can Count On* WITH A BLACK HEREFORD

- Black hair color is the entry requirement for many value-based marketing programs. According to the USDA predominantly black-hided cattle doubled from 1995 to 2010 because of market premiums and branded programs. Today, two out of three fed cattle are black-hided. This is projected to increase to 75 percent by 2015.[3]

- Produce black baldie calves which tend to top local sales across the country while eliminating red baldies off the same set of cows which are almost always sorted off and sold at a discount.

He will be discounted $5.00 to $7.00 per cwt. due to color.

- Black Herefords tend to have black pigment around the eyes as well as dark colored eyes which reduces the onset of pinkeye compared to lighter colored cattle.

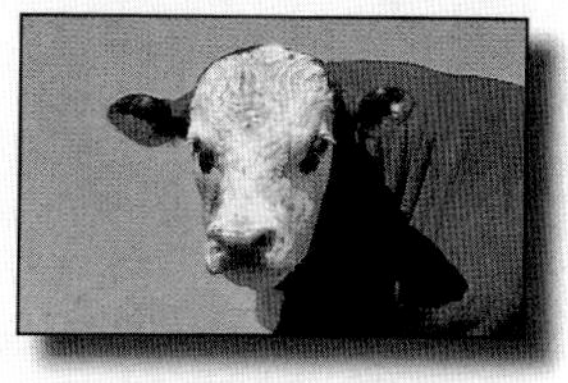

- Black Hereford females tend to have teats and udders with black pigmentation eliminating worries about sun and/or snow burned udders.

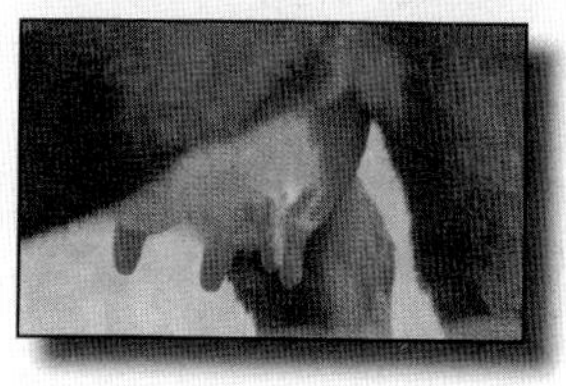

3 Mark McCully, Drovers Journal, July 2011

Black Hereford talking points from the Drover's Journal, *July 2011.*

It sounds so easy, looking back and writing about it. It was not. As Joe often says, "If you think about starting a new cattle breed, here's my advice: Take two aspirin, go to bed, and get over it."

In 2005, only J&N Ranch and Steve and Karen Burwell of Waxahachie, Texas, had registered Black Herefords. By 2008, there were roughly a dozen members and 250 head of cattle in the ABHA registry, with 162 of those from J&N Ranch.

We were committed and determined. By 2012, we had ninety-one members and held the first ABHA National Sale at our J&N Ranch. We scrubbed our sale barn clean and put up cattle panels for our new members to individually display their cattle. It was a kind of exciting brotherhood, starting up something so new. The excitement and enthusiasm our members felt to be the "first" in their state to breed Black Herefords was wonderful to see.

First ABHA National Sale ad, Black Hereford Journal, *Fall 2012.*

ABHA National Sales were held at our J&N Ranch in 2012 and 2013. We then moved to a neutral place, to move the emphasis away from our own ranch. Emporia, Kansas, seemed like a central Midwest location and the National Sale was held there for two years.

I enjoyed organizing these early events, setting up a roster of members, sending out notices, minutes, and agendas for meetings. Getting to know these great people on ranches all over the USA was so much fun. We ordered briefcases with the official logo imprinted on them and would send one to each new member, along with a personalized binder for their registration papers. Yoshi Sato, of our Racket office, designed the original logo for ABHA. Yoshi went on to become the head designer for Garmin, the Global Positioning System (GPS) company.

Some of our early members were new to the registered cattle business and

Black Hereford logo, designed by Racket's head designer, Yoshi Sato.

Joe and Yoshi Sato at the Garmin headquarters, where Yoshi is the head designer.

had difficulty with the requirements of so many weights and measures, pedigrees, forms, and deadlines. Many had never weighed an individual animal before. Weights had to be taken at specific times. They had to buy a calf scale to weigh newborn calves and a squeeze chute scale for larger animals so they could submit birth, weaning, and yearling weights. Many didn't have an ear tattoo kit to put individual identification numbers in a calf's ear. These were the backbone of a registered breed, but these new members hadn't had to comply with regulations like these before.

Running the office and answering the phone was how my days were spent. I truly enjoyed getting to know our fellow cattle breeders. When new members would call with questions it took time and patience, but I did my best to be helpful. Two funny examples do pop into my mind about these difficulties for new members.

Each individual animal was required to have a unique tattoo number put into one or both ears as a permanent identification to match that animal's registration papers. I had one breeder who called and questioned me about tattoos, wanting to skip this step. After explaining everything thoroughly, he said he understood. When his papers arrived by mail at the office, he had tattooed *all* his animals with the identical number! So much for individual identification! I must not have explained quite well enough.

Another member had left her weaning weights off her applications for heifer registrations. I called and explained we needed individual weaning weights to register her heifers. I told her how important weaning weights were in order to evaluate how efficient her mother cows were. She said, "Oh, well then, I have weaning weights. We put all the heifers in a truck, weighed the truck and divided by how many heifers were in the truck." No, I had to break it to her gently, that wouldn't suffice for performance data in a registered cattle association.

We needed a vehicle for education for these members who had new skills to learn about being in the registered cattle business. At our annual meetings we would have different experts give presentations about these new skills and answer questions. Joe contacted many of the animal scientists and professionals we had worked with over the years, asking them to come and give presentations. They would explain EPDs, how to take measurements, formulate effective feed rations, how to take good bull photos, etc.

J&N Ranch had held production sales every year at our ranch since 1986, but it was in 2003 that we sold our first Black Hereford bulls. Our auctioneer from our very first sale was Stanley Stout, an infamous character from the Flint Hills of Kansas. At Kansas State University, there is a Stanley Stout Sale Center, aptly named in his honor. At the beginning of our first Black Hereford bull sale in 2003, Stanley said "Joe Hoagland thinks outside the box." Most people in the cattle business have stories about Stanley. We are no exception.

Our eldest daughter, Anne, at an early age was no different from most farm kids. She would drive around on our ranch, on the inner ranch roads, to do chores or run errands, starting at about the age of twelve. As she was taking something to the sale barn from our house, she was driving in a pickup on the inner ranch road when Stanley Stout pulled into the ranch entrance. Now, Stanley had a red flashing light he could magnetically pop onto his car's roof, and he had a rather lifelike Sheriff badge. He used both on Anne and pulled her over—on our own driveway! Anne still recalls (with heat!) how she tried to defend herself against this "officer of the law." That was Stanley.

An early ABHA annual meeting.

Getting ready for another annual production sale at J&N Ranch.

The cattle industry lost Stanley Stout at too young an age to a heart attack. It was an untimely death that shook so many and left us to find a new auctioneer.

There is a silver lining in even the darkest days. Tommy Barnes of Alabama became our next auctioneer, our friend and adviser. Vickie Bishop had suggested Tommy to us, another thing to thank her for. A true Southern gentleman, Tommy had a way with an auction crowd, softer than Stanley, but exceptionally effective. Always impeccable in jacket and tie, even on the coldest February sale days, Tommy helped us with our twice-yearly sales until he retired in 2023. We always valued the excellent advice we got from Tommy.

Always thinking ahead, Joe used his experience with media and reporters to get our story of this new breed out for the public to see. More members joined due to Joe's advertising work. But to become a commercially viable breed, we knew we had to get more data for comparisons on our cattle. Scientific proof of the superiority of Black Herefords was essential for the breed to survive. We had to appeal to professional cattle breeders, not just "hobby breeders" who enjoyed the novelty but weren't thinking long term. We remembered Vickie's warning.

CHAPTER TEN

BLACK HEREFORD BREED EXPANDS

Membership numbers were growing in ABHA, thanks to Joe's work in advertising and doing great interviews for cattle industry publications. However, from our time in the AHA, we knew we needed to appeal to commercial cattle producers for the ABHA to have a viable future.

Organizing shows and events for our members was fun, but we were looking ahead. Perhaps we were looking too far ahead for some of our members, as there was pushback for programs we began to suggest, such as Whole Herd Reporting. There was continued resistance for more weights and measures that we were asking our members to perform. Again, Vickie's warning of novelty and hobby breeders was remembered.

Every new cattle breed must go through a start-up phase. To reach the next level and prove it is sustainable, it must provide value to the beef supply chain, like some useful traits that are not currently available from existing breeds. We knew Black Herefords had this extra value; we just had to get the word out.

Hereford cattle are known for adaptability in various geographic areas. They tend to be more docile and more feed efficient than most breeds. Animal scientists have said the docility of Herefords aids in better feed conversion—more beef from less feed (See Appendix C). Herefords are not as nervous as other cattle and therefore burn less calories than more high-strung cattle.

Our vision of Black Herefords sought to preserve those basic Hereford traits of feed efficiency and easy handling, with a bonus of more market appeal as the black-hided animals qualified for market premiums. A further genetic surprise was waiting for us, however.

We started by using Angus genetics to turn Hereford hair color from red to black. In our case, black hair color is the dominant gene over red hair color. So, making this change was relatively easy. This was just what Warren Gammon had done in Iowa, selecting Herefords without horns to create the Polled Hereford breed. The polled or hornless gene is also dominant over the horned gene. That makes the transition much easier.

The surprise was, we didn't realize that we would improve calving ease on traditional Herefords. To explain: Angus breeders had for years selected cattle for a big spread between birth weight and weaning weight. They wanted Angus cattle to be small at birth and grow fast while on the cow for a higher weight at weaning. Calves born earlier had a slightly smaller birth weight. What they effectively did was select for a shorter gestation period. This trait came along with the shot of Angus genetics that we put in Black Herefords.

From the very beginning, we hit a home run with superior calving ease. Black Herefords carried this shorter gestation period, with slightly smaller, but vigorous, calves. Getting a live calf on the ground without assistance is "job one" to the commercial cow-calf producer.

Since Joe worked as the director of international activities for the Hereford Association, in 1988 we attended the Tenth World Hereford Conference in England. Prior to that, Joe had traveled to Asia and the Pacific Rim, promoting Herefords, so he knew how to get our foot in the door with Black Herefords in those countries. That experience helped Black Herefords to be recognized worldwide.

In 2011, Joe and I traveled to Australia and New Zealand to promote the new breed. That led to selling bulls and semen into both countries a few years later and the research that showed this superior calving ease trait for the dairy industry, too.

That calving ease was the bonus trait Black Herefords got, besides black hair color, for a marketing advantage. It turned out this tendency for calving ease was unintentional and unknown until research was done in New Zealand dairy herds. In a "beef on dairy" application, using Black Hereford semen on dairy cows, these statistics began to show up, and the data was sent to us.

Black Herefords have a five-day shorter gestation period than traditional Herefords. Calves are born sooner and lighter, improving calving ease. But those calves kept their vigor.

We in the beef cattle business appreciate the calving ease. Nothing is more important than getting a live calf. But for the dairy industry this shorter gestation period was even more important. Milk producers could use Black Hereford semen and get five more days of milk—a production bonus. That is why semen distributors in New Zealand and Australia bought more bulls from us to use.

Thinking globally and traveling to spread word of the Black Hereford breed cost us money personally but helped the new Black Hereford breed find this outstanding breed trait. Years later, Joe and our son Dirck would travel to Brazil to do the same thing.

We might have succeeded with Black Herefords, just based on hair color alone. Qualifying for market premiums based on black hair color was (and is) very real. But couple that with improved calving ease, and the breed grew fast and was quickly accepted by the commercial cattle industry.

Being able to use our J&N cowherd's generations of performance data to measure these important genetic traits that add real economic value to the beef industry really jump-started the Black Hereford breed.

But let's return to the early days and the challenge of getting accurate performance data collection set up and accepted by our members.

Centric Software was working well for our pedigrees and data entry for each animal. Jerry Young, the designer of the program, was able to adjust the computer program to show the table of EPDs for each animal, but we needed help to calculate the formulas for each EPD.

We needed an expert.

In our first Black Hereford sale in 2003, there were no genetic evaluation tools available. We had no EPDs like we had at the time for our registered Hereford and registered Angus cattle. In the first Black Hereford Sale Catalog we only listed the animal's birth weight, adjusted weaning weight and adjusted yearling weight. We handed out ratios showing where the animals ranked within our herd at the sale. But for the most part, purchase decisions were based on these weights and the physical appearance of the animals, with not much else to go on. The ABHA was still a privately held corporation with only one owner, J&N Ranch.

It was not until our sale in 2006 that we had EPDs to explain the differences in the cattle. During the summer of 2005, we hired Dr. Dan Moser, a beef geneticist at Kansas State University, to calculate the EPDs on the few dozen head of Black Herefords registered in the ABHA.

The first challenge for Black Hereford EPDs was to make them comparable to Hereford EPDs. They are predominately Hereford genetics. Hereford EPDs go back to the baseline year of 1976 when they all started at zero. Over the years, Hereford performance improved, so a plus-forty Weaning Weight EPD means this animal's progeny should weigh forty pounds more at weaning than the average Hereford did in 1976.

The problem was, Black Herefords did not exist in 1976, so their baseline year would be 2005 when these first EPDs were calculated. The average Black Hereford numbers for that year would be zero, the starting point. These low numbers would be a marketing disadvantage for Black Herefords compared to traditional Herefords.

Professor Moser came up with the idea of using the top twenty Herefords which had the most proven genetics and average them together to produce a baseline EPD for Birth Weight (BW), Weaning Weight (WW), Yearling Weight (YW), and Milk (M). Black Herefords would be either higher or lower than that average baseline. This made Black Hereford EPDs comparable to Hereford EPDs.

Although Dr. Dan Moser of Kansas State University did our first EPDs, soon after that, he left KSU to become the in-house head of genetics for the American Angus Association.

We next hired Professor David Riley of Texas A&M. David was a beef geneticist and helped us for many years. At that time, we only calculated the basics: EPDs for Birth Weight, Weaning Weight, Yearling Weight, and Milk.

In 2011, a neighbor and longtime friend of ours at our Flint Hills Ranch was elected President of the AHA. Dave Breiner ranched near us in Wabaunsee County, Kansas. We own the Gunbarrel Ranch, one of the larger summer grazing ranches in the Kansas Flint Hills. It consists of all native grasses and barbed wire fence. There is so much rock you can't put up a bale of hay anywhere on its over five thousand acres. Dave could grow winter feed on his place, so we would winter our cows with him on occasion.

Joe asked Dave to try again to see if the Hereford Association would let us join them, now that we had members in several states. Dave tried. The AHA board turned us down again, but this time by only one vote.

It was time to strike out and grow on our own as an association. It was time to move out of our barn office into downtown Kansas City. Time to take another brave step.

Kansas City is in many ways a "Cowtown" and still home to much of the beef industry. From its early days as a beef packing center at the end of long cattle drives, Kansas City today is the home of most cattle breed associations and beef publications. We were outgrowing the small office in our sale barn and the association needed to be more than just about J&N Ranch, its largest breeder.

We decided to move into office space in Kansas City, Missouri. Joe wanted a high visibility location, and the search committee of ABHA board members selected a storefront on a main street in the city center. There it would be seen by industry leaders and visitors to the American Royal Livestock Show, one of the nation's largest such events. Joe was on their board of governors.

To run the growing ABHA, Joe and I hired Dr. Craig Ludwig, a veteran cattle breed association executive. Craig began his career at the Hereford Association as their director of breed improvement—just the job we needed help with! Craig served as the CEO of the Braunvieh Association, also based in Kansas City. Since ABHA had no money to pay Craig's salary, we agreed to pay his salary, and eventually his retirement plan, out of our own pockets for two years and the association would pay us back over fifteen years at 2.8 percent interest. We figured after two years of professional management, the ABHA would be

New ABHA headquarters, opened in 2012, in downtown Kansas City, Missouri.

able to generate enough money to take over his salary and benefits from there.

When the new office opened, we wanted our members to see the new, more professional office. We held an open house to celebrate and invite everyone in the beef industry working in Kansas City.

One of those attending was H. H. "Hop" Dickinson, the former CEO of the American Hereford Association. Hop retired in 1997 but was still a dynamic force in the industry. He was impressed by the association headquarters and the work of his old friend, Craig Ludwig. Years before, in 1985, Hop hired Joe to be the director of international activities for the Hereford Association. We had been friends and had attended cattle events with Hop and his wife, Polly, for years. In a way, our open house was like old home week.

As Joe and Hop talked, he said, "Joe, you need to start a magazine to support your breeders and to help them sell cattle to other breeders." So, Joe took this advice, and another business began for us. The magazine was a popular addition from the very beginning. After ten years of designing our own sale catalogs, we hired a professional. We then used that designer, Gail Lombardino, to help format the magazine. Gail came with lots of experience from working with the Angus Association and others. Joe handled the advertising himself, and we again put up the money to get the magazine started.

We joined the Livestock Publication Council (LPC) and attended their annual meetings for several years. All the major cattle publications attended. Being a part of the LPC with the *Black Hereford Journal* helped promote the breed within the beef industry.

At the grand opening of the ABHA offices. Joe is on the left; Craig Ludwig on the right with Kentucky breeders Wayne Flynn and Tim Tarter.

Many of our members advertised in the new magazine, the *Black Hereford Journal*. It was a quarterly magazine, with lots of information for our new members. Joe contacted animal scientists from Kansas State, Texas A&M, Cornell, and other universities to write articles. Cattle folks Joe and I had met in the past forty-plus years were great contacts for these articles. Members donated photos, Gail helped them with their ranch ads; it was a great timely addition to the ABHA, thanks to Hop Dickinson's suggestion. The magazine never broke even, but that wasn't the point. It was a service to and for our members.

The Black Hereford magazine continues to this day.

As I reflect and record this history, I know we survived the first challenge of creating a new cattle breed with the timing of when this developed in our personal lives. There were several times it could have stalled out and never happened. If the terrorist attack on September 11, 2001, had not stopped Joe's career in professional baseball, he might have pursued becoming a general manager in Major League Baseball instead. If we had continued our life in politics, the ABHA would not have developed. This challenge of timing had been met, but the second big challenge came from within the new association itself. Another Black Hereford Association was created to rival the ABHA.

After moving the headquarters into Kansas City, the breed began to grow

First issue of the Black Hereford Journal, Fall of 2012.

rapidly. America's Heartland filmed a segment on Black Herefords that aired on PBS. Big money from the Texas oil industry came into the breed to raise Black Herefords on their ranches outside Dallas.

That same big Texas oil money created the first rift inside our association. One of the Texans approached us and said, "I want to buy the ABHA." We explained it was a not-for-profit corporation, owned by the members, and we couldn't sell it. This man was clearly not used to hearing the word, "No." He soon started his own association with few rules and poor-quality cattle, but he provided fancy events with dinners and country music bands. Some of our members fell for the sparkle rather than the plain, hard work of raising superior cattle. There is an old cowboy saying for this type of guy—"big hat, no cattle."

At the same time, we had started to attract professional seedstock producers from other breeds, mainly Hereford and Angus, but also Shorthorn and Gelbvieh. These people knew the registered business and began to replace the novelty and hobby breeders that got the breed started.

Divisions in our association grew over tensions between the hobby breeders, the more professional breeders, our CEO, Dr. Craig Ludwig, and the Texas oil money flowing into the breed. Craig had teamed with the Texas oil money, who offered him more pay, to help set up the World Black Hereford Association, our rival. Before doing that, he and the existing ABHA treasurer had paid himself a salary bonus, office expenses, and drained the ABHA bank account down to $300 before they left.

The board voted to fire Craig. They elected Tim Tarter of Nancy, Kentucky, as the new leader. With our support, Tim moved the association headquarters to Bowling Green, Kentucky, his home state, where he was able to get a grant from the State of Kentucky to fund ABHA again until it could get back in sound financial condition.

Another key person at that time, and a future ABHA board member, was Chad Jones of Osawatomie, Kansas. Chad took all the records and pedigrees from Craig's office to his own home for safekeeping. He later sent the rescued records on to the new ABHA office in Bowling Green, with Tim Tarter and Charles Crow of Alabama, who later would also become a member of the ABHA Board.

The Texas-financed World Black Hereford Association was dominated by novelty and hobby breeders who relaxed their breeding protocol and soon lost any interest they had from the commercial beef industry. Within two or three years, they went out of business.

A professional seedstock producer, Tim Tarter had been a longtime registered Angus breeder. As Joe has always said of Tim, "He was the perfect guy at that point in time." Tim attended Beef Improvement Federation meetings to see where the industry was headed. It was there he found Digital Beef, an online registration program. This was a great move to streamline the office. Instead of filling out paper forms which had to be mailed to the office to register calves each year, Digital Beef allowed members to set up an account and fill out their registrations themselves, online. Of course all had to conform with ABHA rules, but this computerization saved a huge amount of clerical work for the association.

At the same time as the move to Kentucky, we sold the magazine to the association, where it began to be run by a professional livestock magazine publishing house.

Both the online registration system and a professionally produced magazine were great steps forward for ABHA, thanks to Tim Tarter's leadership.

Another of the experienced cattle breeders who joined the ABHA was a Hereford breeder, Brock Johansen, of Utah. He played a key but controversial role in advancing Black Herefords as a commercially viable cattle breed, rather than an amateur, hobby breed.

About the time Brock was elected to the ABHA Board of Directors, there was an outbreak of lethal genetic defects discovered in Angus cattle. Both were recessive genes but could become symptomatic 25 percent of the time if both the sire and dam were carriers of the recessive gene defect. One was AM (curly calf syndrome) and the other was CA (fawn calf syndrome). In Hereford cattle the recessive defect was IE (idiopathic epilepsy).

Sensing the ABHA could become a dumping ground for cattle carrying these recessive genetic defects, not visible to the eye, but likely to appear in their offspring, Brock insisted on genetic testing for these defects in Black Herefords. This began the "Certified Herd Sire" label for ABHA. Only bulls that have been tested free of these genetic defects can be used to produce a registered Black Hereford, according to this rule.

It was the correct decision. Like Tim Tarter, Brock saved the breed at a crucial time in its' history. The novelty and hobby breeders opposed these additional rules. At the 2013 board meeting, a small breeder from Illinois said she was "concerned about too much regulation" coming from the ABHA office. She wasn't thinking long term. Our larger breeders, like Brock, also perhaps understood the threat these genetic defects posed and were more willing to undergo herd genetic testing to keep our Black Herefords free and clean genetically. Small hobby breeders didn't want all this testing and recordkeeping.

The best safeguard to this type of narrow attitude was a provision in the association bylaws that was suggested to us, early on, by Dr. Bill Abel, a family friend and CEO of the American Charolais Association. He said to "base voting in your association membership on the number of cattle registered by a member and not based on one man, one vote." He correctly observed that larger breeders pay more of the association expenses than small breeders and should have a greater say. ABHA had been set up with voting that way, but there had been a push by the smaller breeders to change, giving them more control. This was another time of contention within the breed. We were able to keep the voting based on calf registration numbers, as had been originally set in our rules and regulations. The vote to change the rule was never close.

ABHA was making good progress. We were getting more of the members to send in performance data to expand and make our EPDs more accurate. However, Joe knew we needed to have EPDs compared directly with other beef breeds to attract more commercial ranchers to choose Black Hereford bulls. A rancher needed to be able to directly compare the performance of a bull from an Angus, a Hereford, a Simmental, or any other breed to a Black Hereford bull. We knew our bulls were of equal quality and performance, but our EPDs were only within our breed and only for four basic traits.

Through the years, as we got to know more people within the beef industry, we learned about International Genetic Solutions, or IGS. Here was our answer to more complete EPDs. IGS was a multi-breed genetic evaluation system started by the American Simmental Association under the leadership of its CEO, Wade Shafer.

Together, Richard Spader of the American Angus Association, who started Certified Angus Beef, and Wade Shafer have impacted beef production more than anyone else in our lifetimes. Wade developed IGS to evaluate cattle from multiple breeds into a single framework. Commercial cow-calf producers could look at one set of numbers and compare those to the same genetic evaluation in another breed. This tool was what we needed to show Black Herefords as the best choice for America's commercial cattle producers.

Meanwhile, Dr. David Riley of Texas A&M University continued to calculate EPDs for the ABHA.

However, starting in 2015, we enrolled our J&N cowherd with IGS and then began to use their EPDs to evaluate our J&N herd. We used IGS data in our advertising and sale catalogs. This expanded the number of measured traits to include calving ease (CE), maternal calving ease (MCE), and carcass traits like yield grade (YG), marbling (MARB), and ribeye area (REA). We also added selection indexes, all-purpose index (API) and terminal index (TI).

About this time, J&N's Black Hereford cowherd attracted cattle researchers. In 2016, Dr. Riley led a study involving our cowherd to locate the DNA marker for eye pigment in white-faced cattle. Another research project on our ranch was done by the University of Leuven in Belgium, under the direction of Dr. Steven Janssens, looking at a variety of maternal traits.

After serving a number of years as president of ABHA, Tim Tarter and his wife Teresa had been running the association largely by themselves. The ABHA Board decided to again hire a professional CEO. They hired Ernie Dominguez. Ernie lived in Texas, however by this time, our need for a brick-and-mortar office was becoming obsolete. The association could be managed virtually, utilizing online and remote services.

Rod Garman, a longtime breeder from Oklahoma was elected as the new president of ABHA. Again, he was the right choice at the right time.

At the National Cattlemen's Beef Association conference in Houston in 2022, Joe and I hosted a dinner in a private room at a local restaurant. We invited Wade Shafer and several principals of IGS who had been working with J&N cattle for seven years to meet with Rod Garman and Ernie Dominguez. Joe and I wanted to be sure Rod and Ernie had a quiet chance to hear what IGS could offer to Black Herefords, with their multi-breed EPD comparisons.

Rod Garman understood immediately and led the association to the next level. As Tim Tarter had done, creating an online registration system, Rod advanced our genetic evaluation system by joining the largest cattle performance database of IGS. He also got Black Herefords listed on the Oklahoma State University website of cattle breeds. They were the last holdout to recognize us as a separate cattle breed and something other than a crossbreed.

Becoming a member of IGS probably did more to advance Black Herefords as a viable commercial cattle breed than any other single event. Joe gives credit to Rod Garman for leading ABHA to join, saying, "Rod made a huge contribution to Black Herefords by moving us into the multi-breed EPDs with IGS. That gave us great credibility with all bull buyers that they could compare our EPDs against all these other breeds."

By 2025, the ABHA was the fastest growing cattle breed association in America. Today there are over six hundred active breeders in over forty states, Canada, Mexico, New Zealand, Australia, and Argentina. All of them trace their origins back to J&N Ranch.

Today, our breed has no interest in being a part of AHA, as we had in the past. Unlike other beef breeds which are declining in annual calf registrations, Black Herefords are increasing. Our EPDs are better, and our financial position is strong. The breed has arrived.

CHAPTER ELEVEN

OUR EUROPEAN LIFE

After the terrorist attack on September 11, 2001, our airline business nearly stopped in the USA. Long security lines and early airport arrivals ahead of flight times resulted in a surge of dining options in the terminals themselves. Less food was served on domestic flights so there was less need for the passenger service products Racket sold. We were selling less in the United States. However, we had many contracts with European airlines. As the European Union consolidated as a single market, airlines there looked to fellow European suppliers, not American ones.

In order to continue, or hopefully do more, business in Europe, we needed a European office. We needed a somewhat neutral location, not England, France, or Germany. We settled on Brussels, Belgium, the capital of Europe, and home of the European Union. Brussels is a short distance from London, Paris, and Amsterdam, a great central location in the heart of Europe.

For us as Americans, another helpful benefit of Belgium is that English is spoken frequently. Living and working there are people from all twenty-seven other EU member countries, so English is the common, default language. There are English television stations, which was important to our American employees who had transferred over there. By 2002, we had effectively moved our sales team to Brussels.

As a result of being there for business, Joe and I have lived in Belgium, on and off, for nearly twenty-five years. Two of our children went to school there. Our oldest daughter, Anne, speaks French quite well. She had begun to work at Racket during her college years and continued after graduation. She opened our Brussels office and lived there with her husband Brandon, as a newlywed couple. Brandon, as a trailing spouse, was unable to get a work permit in Belgium, so he decided to apply to Cambridge University for a post-graduate program. He was accepted and earned a master's degree in international food safety and disease traceability, while commuting to England from Brussels. Brandon's degree would prove helpful in another business we would start together—CattleTracs. Later, our son Dirck worked in our Brussels office also, while getting his MBA from the Solvay School of Business and Economics. It is considered one of Europe's top MBA programs with students from all over the world.

Belgium is a wonderful "little kingdom" as it is affectionately known. Although bureaucratic at times, Belgium takes good care of its citizens. Healthcare is excellent, ranked third in the world, with a single-payer health care system. Yes, taxes are higher than in the US, but the costs of doctor visits, medication, and physical therapy are a fraction of the price they are in the US.

Two other remarkable differences are: Belgian universities are free. If a student's test scores are in an acceptable range, four years of university are free. Banking transactions are very secure, using pin codes. No one writes paper checks and there is not an identity-theft problem there.

Also, Belgians still remember the help America gave them in WWI and WWII, believe it or not. Especially during WWI, Herbert Hoover, before he was president of the USA, arranged food aid around German blockades. Belgians still honor him as having saved their children. They remember the help of America; you can easily see when you attend Memorial Day services at any of the many American military cemeteries throughout Belgium. Decorating a service member's gravesite is an honor passed down in families from generation to generation. The word "occupation" means something different to a Belgian—not what you do for a living. We feel fortunate to have a chance to experience that respect for our country. It makes us proud to be American.

But how did we get involved in the agricultural community in Europe? We've been asked this many times! This has been a source of information and enjoyment for us both, especially as Joe retired from the airline business, and we have had more time to pursue our ag interests. Here's how it began:

In 2014, Joe was serving on the tax committee for the National Cattlemen's Beef Association (NCBA). They were discussing inheritance taxes, as landowners always are concerned with being "land rich and cash poor" and how to pass valuable farmland on to the next generation. Since Joe was in Europe frequently on business, it was suggested that he research how farmers and ranchers in the EU handled the topic of inheritance. This was just the kind of challenge Joe was ready for, with the ABHA on firm footing and with our two older children beginning to work in our businesses.

Joe had used his membership in the Livestock Publication Council and as publisher of the Black Hereford Journal to join the Press Club of Brussels. He got a press pass for the European Parliament and the European Commission. It was there that he met Thierry de l'Escaille. Thierry was a gifted lobbyist representing the farmers and landowners in the European Union.

The European Landowners Association (ELO) office welcomed Joe to come in for a chat. Thierry de l'Escaille was the CEO of this group of agricultural businesses who combined their efforts to lobby for landowner friendly policy at the EU. ELO also has an organization for the individual landowners themselves, named Friends of the Countryside (FCS). The major landowners in Europe are largely the old, aristocratic families who have held and cared for their land for generations. They make up the membership of FCS. As Joe says, some have money, some have titles, some have land, but few have all three.

After meeting several times and inviting Joe to join in some meetings, Thierry suggested we join the Friends to get to know other agricultural families in Europe. In addition to meetings in Brussels, FCS organizes an agricultural tour to a different country each summer. We thought this would be a great way to understand more about different types of agriculture and to bring back information to the USA. We had no idea how much fun would be involved!

Being the only American members of FCS, we were at first a bit timid at these meetings but were embraced wholeheartedly by the membership. I was in awe of how much these citizens of Belgium, France, Germany, England, Poland, Bulgaria, Italy, knew about American politics and events. We heard without reservation how much America is watched by the rest of the world and how important our place on the planet is.

Slowly we developed social connections in Europe. Joe and I thoroughly enjoyed each tour in a different country and made friends with the members of FCS. We were invited to grand parties held by old aristocratic families. First in Belgium, then Switzerland, Germany, France, Spain, Portugal, Poland, Italy, and Sweden. These noble families across Europe were landowners, and through the generations, many knew each other. They all shared an interest in the CAP (Common Agricultural Policy) in the European Union. Farmers in Europe are valued more highly than they are in the USA. Food security is equated with national security in these countries that have been occupied at some point in their history by invading forces. Support for food production in Europe was the EU's largest single budget item until the reduction of the USA monetary support for NATO, starting in 2024. Now more is being spent on defense than agricultural subsidies for the first time since the EU's founding.

These counts, countesses, barons, baronesses, princes and princesses were curious about American landowners. I have to admit that curiosity went both ways. At our very first Friends of the Countryside meetings held in the Netherlands, there was a formal dinner. I found the tradition was to separate couples so that you each got to know different members. I was seated next to a prince! It took all of my confidence, bundled with nerves and fortified by my years of experience, to calmly be seated and begin a conversation with this dignified, quiet man. I was so fortunate, looking back. He was a kind soul, and we shared this, our first of many conversations about education, philosophy, and mankind. I look back on that with a smile and think—there I was, a girl from Tecumseh, Kansas, having dinner with a prince! We were shown kindness and yes, curiosity, and had so many meaningful conversations with our European Friends over the years.

We arranged a study tour in America for many of these Europeans to visit our friends Jamie and Judy Clement, descendants of Captain King, founder of the King Ranch in Texas.

On another trip to the USA, members of the Friends of the Countryside

Joe and French Count Ghislain de Nicolay at the King Ranch, Kingsville, Texas.

were invited by Tony Simmons and his family to Avery Island in Louisiana, for an exceptional tour. Avery Island is home of Tabasco and still a family owned and operated business. The Tabasco brand is known throughout Europe and the world. Our European friends were fascinated not only by the process of producing pepper sauce, but how the family had maintained control within the family of a worldwide business for so many years.

We also found so many Europeans have a personal or business connection to America. Many had come to American universities for their higher education. One German friend had spent many summers as a child at a farm owned

Tony Simmons, past president of Tabasco, at left, with German Count Gunther von der Schulenburg and Emilio Navarro-Martinez of Spain, during a visit to Tabasco's headquarters on Avery Island, Louisiana.

by his father near Sikeston, Missouri. Another, Tilman Behrens, a Canadian German, had grown up in Canada but had spent so much time in California that a childhood friend and next-door neighbor was Gavin Newsom, the current governor of California. Many of our European friends spoke with such affection and enthusiasm about their families' long vacations in the USA and how friendly Americans had been to them as travelers. It was heartwarming to hear these stories.

I can't even begin to cover all we learned about different types of agriculture during these summer tours with the Friends of the Countryside. We shared stories about difficulties: weather, crops, pests, governmental regulations, reintroduction of wolves, etc. We visited vineyards, greenhouses in the Netherlands, a dairy in Poland, saw experiments with biowaste to produce power. Olive oil production in Italy, rose oil distillation in Bulgaria, cork harvesting in Portugal—so much to learn and see!

Cork farming in Portugal—I'd never imagined this. It takes several years for the cork bark to form adequately to then be harvested, so the year for next harvest is painted on each tree's trunk.

Spain's Iberian pigs, famous in Europe for their mild flavor, roamed large pastures. Their secret was a diet largely of native acorns from trees in those pastures. This was so different from the confined pig operations I had seen in America. Such happy pigs!

Rose oil production is famous in Bulgaria—the very best quality worldwide. Near Plovdiv, in the best valley for roses, our lovely young friend explained how the roses are hand-picked and distilled into rose oil. Their high quality rose oil is purchased by top cosmetic firms, like Lancome and Chanel. Concerns about finding labor to continue rose oil production has led them to expand the family's agricultural interests into vineyards.

One of my favorite stories about comparing American to European agriculture came in an unexpected place. This was a ranch tour in Spain where bulls are raised for bull fights. I've never been to a bull fight but know the basic concept. Their office had magnificent photos of their prize-winning bulls and matadors and trophies galore.

We were invited to get into a wagon and drive out into the pastures where the younger bulls are raised together in pastures. We were admonished to keep quiet, as any disturbance could get the bulls excited and they would begin to fight and might harm each other. We didn't want this to happen—these ranchers were kind enough to let us tour! As a beef breeder, I wasn't impressed by the bulls' phenotype, but these weren't raised for steaks.

Once we were back at their office, I asked the owner if they used a software program to track performance data as we do in the USA. He pulled out his cellphone and showed me the app. It was very similar to the evaluations we track for Black Herefords. There was one huge difference, however. They select for aggressive temperament in their mother cows, which made sense. And naturally, we cull for any slight tendency toward aggression. Exact opposites—and so very interesting!

We found the concept of private land ownership was different in Europe. Public access to private land is a way to get favorable tax incentives. You could build a walking trail through your land and allow people to hike, picnic, and enjoy nature. This is hard to imagine working in America. But from this experience, we decided to try a kind of public access on one of our ranches, leasing it to the State of Kansas for walk-in hunting. It has worked quite well and has increased our ranch income in a new way.

Some European estates were having a difficult time financially, so we saw what kinds of additional businesses they had introduced or started, to add income to their bottom line. House and garden tours for visitors, olive oil tastings, wine tastings, venues for events and weddings, all were new revenue streams we saw landowners trying. From these conversations, we decided to try some selective logging back at our ranch in Leavenworth. This isn't something we would

Joe and Norma's native black walnut tree harvest in 2023.

have tried without the experience of our friends in forestry in Europe. Here was another source of income for landowners.

Also, we found that every family had its own unique story and troubles in being profitable and keeping the business in the family. People in agriculture are people, no matter where they live.

One German prince was especially kind and friendly to "we Americans." Having vineyards on land that had been granted to his family by the Holy Roman Emperor in 1147, he told stories we can scarcely imagine. Their land near the Rhine River had been taken away by war or politics seven times. It had been in Germany, then France, then Germany again. They had to acquire parts of their land over and over again—but had successfully done so. His two sons were

Joe and Prince Michael Salm of Wallhausen, Germany, in front of the Salm Castle which was destroyed by Napoleon.

in the family business now, and it would continue. They are the thirty-third generation of the family on that land.

Another German friend told the story of his grandfather leaving their farm when Hitler took over Germany. The Communists had seized most of the land in Eastern Germany from 1945 to 1949. Luckily, his grandfather kept his land deeds, and when the Berlin Wall came down, he was able to show ownership. He was then allowed to buy back the family land that had been in East Germany for years. Although it was a financial burden, it was in a way a triumph for his family to have that land again, and a source of pride for our friend.

But not every family had that triumph. At a meeting held in Warsaw, one of the members of the Friends of the Countryside asked if I'd like to take a walk with her and see her grandfather's house. Of course, I said yes! As we walked along the main street in the old town of Warsaw, she pointed out a church her family had built for the citizens. Across the street was a huge, stately home with garden and gates. This had been her grandfather's home. He had been alerted and fled with his family to Spain just two weeks before the Nazi invasion of

Joe and Norma in front of Torre a Cona, the estate of our friend Count Lorenzo Rossi de Montelera, whose family founded Martini & Rossi vermouth in 1863.

Warsaw. Unable to safely return to Poland for so many years, this magnificent home was now a governmental ministry office. I could only imagine the longing she must have felt, to look through the iron fence at the place her family had, for generations, called home.

Previous ownership of land has been handled differently in each of these countries. Germany only gave land back to Jewish families and those anti-Hitler conspirators whose identification had led to Nazi confiscation of their land. For most others, former owners had to buy back their land from the Federal German Government. Others, like Croatia, were slow to move ownership from the new governments to private individuals. Ukraine had given each citizen a small parcel of land, but these were farmed collectively in large swaths by wealthy oligarchs. Each citizen receives a payment yearly, however, from their small farm acreage crop.

Today, because of these transatlantic friendships and sharing of common interests, we have a network of friends and invitations to visit their farms and estates throughout Europe. May we live long enough to visit them all and repay the kindness we have experienced.

Vineyards, sunflowers, canola oil, wheat, corn, hay: all the same crops as in the USA and each family with their own story, just like in the USA. We even found our experiences in America were of equal interest to Europeans, and I was so pleased to be interviewed for a series of "Women for the Future of Agriculture," published in 2023.

Joe served two terms on the Friends of the Countryside Board of Directors and had access to decision-makers at the highest level on agricultural policy in Europe for eight years.

These have been a fascinating addition to Joe's and my lifelong learning about agriculture. What a privilege to have had this opportunity!

ELO

European Landowners' Organization

WOMEN FOR THE FUTURE OF AGRICULTURE

Norma Hoagland

Ranch manager

I am Norma Hoagland, age 72 from Kansas, USA. My husband Joe and I own and operate two beef cattle ranches and a feedlot in Kansas. We own and operate citrus orchards in the Rio Grande Valley in Texas. We have three children, three grandchildren, and I am an author and biographer.

Your journey

I grew up on a dairy farm in Kansas, on land my family has owned since 1866. On that farm, we raise corn and soybeans. I have always been involved in agriculture. When Joe and I married, we dreamed of running our own beef cattle business. Joe was involved in his businesses in the city, and so I was heavily involved in the daily management of our ranches.

Your profession

My university training was as a teacher, so I researched and learned about animal husbandry, nutrition and grass management. I became our artificial insemination technician. I computerized our cattle pedigree information and performance data, then began to do the same for other breeders as we established our own breed of cattle - Black Herefords.

"Do your job and recognition will follow."

Challenges along the way

In the male-dominated cattle industry, there is always the challenge for a woman to be taken seriously. One personal example was in my classes to get my Artificial Insemination license, I was one of 2 women in the class. We were called out and treated differently than the men. However, the overall challenge in agriculture is profitability, for everyone.

The biggest misconception is that a woman can't possibly do the work or know anything! Customers would call our ranch to ask about our breeding bulls and insist they had to talk to my husband.

I would encourage a woman to forge ahead. This is a particularly interesting time to be in agriculture as food becomes more a focus in our countries. The future may be in land management, as ownership of land is becoming cost-prohibitive, so a woman can channel her interest into managing for others. If you are lucky enough to own your farm or ranch, don't get discouraged or angry at others. Do your job and recognition will follow.

Women in agriculture

During the years I was actively running our ranches, there were not a lot of organizations or projects with other women. We women would visit during cattle seminars and encourage each other. I did a number of children's classroom lessons on agriculture, working with women teachers in my community. Currently, I see there are many more organizations that promote women in agriculture and give them a chance to work together to show the face of agriculture to our urban communities. That is a good thing!

25/03/2023

Norma is featured in an article "Women for the Future of Agriculture" published in Europe, 2023.

CHAPTER TWELVE

CATTLETRACS

In the years we were working to create the Black Hereford cattle breed, other business opportunities arose for us. Because of our early political life, we always watched the news, both domestically and globally. Our children can tell you we discuss current events every night at family dinner.

Joe was ever the entrepreneur, always looking for ideas to improve lives and increase our businesses. Starting the Black Hereford breed led to the publishing of a magazine, the *Black Hereford Journal.* Using the magazine to get press credentials to the European Parliament and membership to the Press Club of Brussels, we kept abreast of newsmakers in Europe. But an even greater opportunity arose in some ways quite by accident. CattleTracs. It is a cellphone app designed to track cattle in the beef supply chain.

During the years we were setting up the ABHA but living in Europe, three months each summer and two months each winter, we were forced to work on Black Hereford issues remotely. None was more pressing than finding a better way to do genetic evaluations of the cattle. We knew we needed to find ways to improve how we determine the genetic value of the cattle we sell others for breeding purposes.

It turned out that one of the world's leading molecular beef geneticists was working at a university in Belgium. His research on Belgian Blue cattle used DNA to locate certain inheritable traits. DNA sequence analysis is today the basis for Genomic Enhanced EPD's which are used in all the major beef breeds and for Black Herefords.

Through those introductory meetings, Joe met one of the students who was working on his PhD at the University of Leuven in Belgium. That student came to Kansas and worked on our ranch to collect data on Black Herefords. One day as Joe went to the student's office in Belgium, he noticed a young Chinese graduate student who had developed a facial recognition algorithm for sheep.

Asking if it would work on horses or cattle, she said yes, it should. That started Joe on another family business venture—because of Black Herefords—and Joe's zest for learning.

Joe hired researchers at the University of Leuven to develop a facial recognition algorithm for horses and cattle. With the help of Brett Spader, son of Richard Spader, the past CEO of the Angus Association, we applied for a business development grant from the State of Missouri.

It was also helpful that our son-in-law, Brandon Bauer, had recently received his master's degree from Cambridge after his thesis on European Food Safety and Traceability. We named Brandon as president of this new corporation. Brandon, with the help of Brett Spader, worked on the application to get the business development grant from the State of Missouri. Then Brandon managed the grant as we developed a cellphone app using facial recognition technology.

As a registered quarter horse breeder, Brett was able to get a contract with the American Quarter Horse Association (AQHA) to test the facial recognition concept on horses as they aged.

The CattleTracs office in 2021, located at 719 Walnut, Kansas City, Missouri.

It had always been an issue for the AQHA to prove a horse in competition (at the age of six or seven) was the same horse that had been registered in the AQHA as a colt. Using Radio Frequency Identification (RFID) chips for proof of identity was never acceptable to horse owners as those chip inserts would migrate in the warm-blooded animals. Ours was a better answer to that problem, but we could never collect enough horse images to train the algorithm. With horses, we could only collect a few dozen images. With cattle in feedlots, we could collect thousands of images. Cattle should be easier, we thought.

With the help of Dr. Dan Thompson, the former head of Animal Science at Iowa State University, we collected enough images of cattle to use machine learning to create a facial recognition algorithm. Dan had been to our ranch in Leavenworth and made a presentation to Black Hereford breeders at one of our annual meetings. That is how we knew him. We sent those images and others

Joe, displaying the CattleTracs app.

back to our researchers at the University of Leuven.

Then we hired Kansas State University to do proof-of-concept research to validate the project. Under the team leadership of Dr. K. C. Olson, it worked. News spread fast, including stories on NPR and CNN as well as reports in Europe.

The World Organization for Animal Health (WOAH) based in Paris, France, endorsed the project to provide disease traceability in cattle. Joe was awarded the innovation prize by the Nebraska Cattlemen's Association.

We hired computer engineers to develop a cellphone app that would autonomously, or automatically, photograph a bovine face. It would not photograph anything else. It would also automatically stamp that photo with the date and GPS location of where the photo was taken. That information was stored in a database, using cryptographic hash functions like ones used for cryptocurrencies. Once recorded in blockchain, the information could not be altered.

This provided a secure database with third-party verification in blockchain to trace the movement of cattle in the beef supply chain all the way to the grocery store. QR codes on beef packaging in the store could access that information, or perhaps you could watch a short video clip to see the rancher who raised the beef and hear a description of their operation.

We did a demonstration for Tyson Foods and discovered the facial recognition algorithm did not work on solid black cattle faces as the cattle grew more hair in winter.

However, the database management system, recording photos with the date and GPS location in blockchain, worked perfectly. Finding a use for this database system for traceability was the next logical step.

CHAPTER THIRTEEN

SAVING THE AMAZON RAINFOREST

In 2020, just before the pandemic started, our son Dirck and Joe went to Brazil as part of a trade mission to promote and sell Black Herefords. They saw firsthand the change in Brazil's agriculture, from vast grasslands for cattle into vast soybean fields. The cattle had moved elsewhere in Brazil.

American agriculture is largely based on two crops: corn and soybeans. The United States produces 32 percent of the world's corn and 28 percent of the world's soybeans. We have 5 percent of the world's population. American farmers need to export grain. Trade tensions with China are challenging this need. China now buys its grain from countries other than the USA.

When the USAID program was terminated, what American policy makers did not understand is that the USAID program was not so much about feeding the poor in Africa. It was about buying up America's grain surpluses, especially wheat and milo, to support American farmers and stabilize grain markets.

Grain prices have dropped as exports to China have slowed, and surplus production formerly used for USAID has been dumped on the domestic market. Prices dropped immediately when USAID was terminated. We will never go back to feeding the world. China has invested too much in Brazil to reverse course and return to buying from the USA. China encouraged Brazilian farmers to grow corn and soybeans, which Dirck and Joe saw firsthand visiting farms that used to raise cattle. In Brazil, they can get a crop of both soybeans and corn every year, off the same field. The growing season is year-round. In the States, it takes us two years to produce that much. China has invested in farm-to-market roads and has built huge on-farm grain storage facilities in Brazil to extend grain marketing throughout the year in Mato Grosso, the state just south of the Amazon.

Immediate results of this lack of exports are visible to us. The Archer Daniels Midland (ADM) grain terminal on the Missouri River, just a mile down-

A once vast grassland where cattle grazed is now a soybean field in Mato Grosso, Brazil, in March of 2020.

stream from our ranch in Leavenworth, used to load barges and trains full of grain for shipment to the port of New Orleans for export. That terminal is closed and is for sale today. Our nearest John Deere dealership in Lawrence, Kansas, has gone out of business.

Much of the pressure on the Amazon comes from soybean plantings in the Mato Grosso, a vast grassland plain area of Brazil that used to be used for cattle grazing. These native grasses were plowed and became fields for soybeans which were shipped to China to replace soybeans that China used to buy from the United States. This was a result of the first trade war against China in 2017.

Grazing cattle had to move somewhere else. The cattle moved north into the Amazon Rainforest, just as quickly as the trees were cut.

Cattle prices are at an all-time high as cattle inventories have dropped not only in the United States but in Europe and Australia as well. This is a result of drought, wildfires, and climate disruptions. The age of American farmers and ranchers is a factor, too. Not every family has a child willing to work hard to stay and keep working the family land. Worldwide demand for beef has remained high, partly due to wealth increases in Asia and the Pacific Rim.

Food grown on trees has outpaced all other foodstuffs in price increases. Olives, almonds, cacao, coffee, and citrus prices have all jumped as supplies shrink from climate change. It takes time to grow new trees that have been damaged or lost to extreme weather events. Some areas will never be replanted again.

We follow air travel more than most families since we are in that business. Wealth, as measured by people who can fly on an airplane for a vacation, has shifted. In the 1970s, 70 percent of all airline passengers worldwide were American or European. Today, 70 percent are from China, India, the Middle East, and South America. As their disposable income increases, so does their demand for beef.

As a family, we feel well positioned with Black Hereford cattle and citrus, but as a country, the USA is not well positioned in global agriculture. We have lost too many of our traditional markets that took America years to develop because of needless trade wars.

Verification of cattle's impact on the Rainforest came through the story of Dom Phillips, as reported by Matthew Hay Brown in the Washington Post. Dom was a British investigative journalist working in the Amazon. He and others concluded that 90 percent of the deforestation of the Amazon Rainforest was done to expand cattle grazing. They were preparing to publish their findings. Dom Phillips and fellow journalist Bruno Pereira were both killed on the front line of the battle to save the Amazon. A murder trial is pending.

As the world began to warm, Europe sought to impose its economic power on other countries to help alleviate global warming. The EU's consumer market of over 550 million people was larger than the USA with its population of 350 million. They passed a law in 2023 that products sold in the EU could not be sourced from recently deforested land. These products included cocoa, coffee, cattle, palm oil, soybeans, wood, rubber, charcoal, and paper. This European Union Deforestation Regulation (EUDR) was initially set to take effect on December 30, 2024.

Enforcement of this regulation was left up to each member state of the EU. They were to each determine if products entering the supply chain complied with the law. Member states were unprepared as the deadline approached. The European Commission recommended a postponement. The Parliament voted to approve the delay 371 to 240 with thirty abstentions. The vote was largely along party lines, with the European Peoples Party prevailing over the Green Party and the left-leaning side of Parliament.

To enforce the EUDR, a Deforestation Due Diligence Registry was created for operators in the supply chain. This electronic registry hopes to identify producers of crops and middlemen to facilitate enforcement by the member states.

Operators can draw areas on a map or provide coordinates where products are grown. Locations previously used can be reused. Traders in the supply chain can reference an operator's previous statements to prove due diligence as to source of products was performed. All of this is done, however, without any third-party verification.

Enforcement of the EUDR remains problematic. It covers too many products, with too many rules, too much paperwork and bureaucracy. It is without any third-party verification that proves what is said is in fact true. These remain huge obstacles.

Our cellphone app, with the date and GPS stamp, which did work well in CattleTracs, could be the solution. To save the Amazon, the EU needs to forget about all these other products which are the cause of much objection from European producers about the EUDR. Focus is needed exclusively on cattle, the true cause of the recent deforestation in Brazil. The EU needs to focus on the Amazon Rainforest. Only the Amazon creates its own weather and serves as the "lungs of the world."

The Congo Basin in Africa is the second largest rainforest on Earth, after the Amazon. Spanning six countries across central Africa, the Congo's forests remain largely intact. The world's third rainforest region is the Borneo-Mekong basin in Southeast Asia. Much smaller than the other two, it is not pressured by deforestation. Neither of these areas have much beef production or pressure to create more areas for cattle grazing, and they are considered less endangered.

The Amazon is massive by comparison to these two and located in one of the largest cattle producing areas of the world. Brazil has well over twice the number of cattle than the USA has. There are no land deeds in the Amazon. If you are in possession of land, it is assumed you are the rightful owner. This leads to a great incentive to burn and clear quickly to start to graze cattle. These cleared grazing areas are getting so large, they are visible by satellite images.

To save the Amazon Rainforest, the world needs to stop buying beef produced in these deforested areas.

Joe proposed to the European Commission to use CattleTracs as a cellphone app that would be free for farmers to use and record the date and GPS location of the cattle they raise. These recordings would be verified to be accurate by using cryptographic hash functions to store data in blockchain so it cannot be altered, just as it works in cryptocurrency transactions. Third-party verification is more important than self-serving statements of producers and suppliers.

A farmer would open an account on the CattleTracs app with his name and address. That farmer and each subsequent handler in the supply chain would answer two questions: Where did you get your cattle? What did you do with them? Those answers with the date and GPS locations would link together in a vertical one up-one down digital ledger to provide traceability. Beef in the grocery store with a QR code on the package could show the source and farmer at his ranch. This QR code would access the digital ledger and show the beef did not come from recently deforested land.

Joe's proposal is pending before the European Parliament and Commission President Ursula von der Leyen.

The lead editorial in the October 25-31, 2025, issue of The Economist, page 10, entitled "Saving Tropical Rainforests—The Lessons from Bazil" clearly makes the case that preventing cattle production in recently deforested areas is vital. As they say in the accompanying article on page 42, "Where there's smoke, there's cows. [. . .] Some 67,000 square kilometers of virgin rainforest were destroyed last year, an area roughly the size of Ireland and nearly twice as much as was cleared in 2023." Joe says they need to forget worrying about where pulp and paper come from and track cattle in the beef supply chain to prevent further deforestation. CattleTracs can prove where cattle are grown.

He is still working toward this goal. And he has an even greater sense of urgency today.

On November 13, 2025, the Press Service of the European Parliament stated that between 1990 and 2020 over 1 billion acres worldwide have been deforested, especially in the Amazon, Africa, and the Pacific Rim. In Brazil, most of the recent deforestation occurred in the Amazon under Bolsonaro's government from 2019 to 2023. His response to Europe's concern about the rainforest was "plant more trees in Europe."

When the trade war first began with China in 2017, China replaced soybean imports from the USA with soybeans from Brazil. With two crops raised in rotational plantings, they reduce the nitrogen needed to fertilize corn with the residue of the soybean crop planted earlier. Soybeans are a legume that releases nitrogen back into the soil which the following corn crop uses as fertilizer. With increased corn production, Brazil will build confined cattle feeding systems to produce beef to rival the quality of American beef.

Joe saw China's economic influence as it has grown from his first visit to China in 1985. The original BRICS countries are Brazil, Russia, India, China, and South Africa. They have begun to challenge the dollar as the world's reserve currency, which is used for international trade. Currently, US dollars are still the dominant world currency, but we see evidence that it is beginning to change. BRICS has expanded to eleven members and now includes Iran, Egypt, Ethiopia, United Arab Emirates (UAE), Indonesia, and Saudi Arabia. The GDP of BRICS is larger than the G7 today.

Russia is the world's largest producer of potassium and phosphorus, both of which are used as soybean fertilizer. Sanctions were put on Russia because of their attack on Ukraine. We found, through a business friend in Sao Paulo, Brazil, that Russia was shipping fertilizer to Dubai, a port in the United Arab Emirates and a member of BRICS, relabeling it, and then shipping to Brazil to avoid sanctions on Russia. Brazil pays Russia with the Chinese Yuan that they get from selling soybeans to China. Russia then buys military hardware from China, especially drones, with the Chinese Yuan they are paid by Brazil to buy fertilizer. All this is done to avoid American sanctions and all done without US dollars.

In the fall of 2025, the Federal Reserve cut interest rates three times. During that time, the yield on US Treasury bills has continued to increase. As I understand it, that never happens. This may be a sign that demand for dollars is softening due to more global trade being done in other currencies.

Trade and agriculture are global. We all need to pay attention to events outside the USA to fully understand what is happening within the USA. This is just evidence that America as the world's leader on global issues is beginning to collapse. Issues like deforestation of the Amazon will be decided not by America, but by other countries who will be replacing American leadership. That's why Joe is working with the European Union to use his cattle tracking system.

On November 25, 2025, the European Parliament voted 402 to 250 with eight abstentions to once again postpone the implementation date for the EUDR. Large operators and traders in the supply chain will have until December 30, 2026, and small farmers and traders will have until June 30, 2027, to comply. The Press Service of the EU Parliament released this statement: "This additional time is intended to guarantee a smooth transition and allow the implementation of measures to strengthen the IT system that operators, traders, and their representatives use to make electronic due diligence statements." CattleTracs is a perfect solution to be one of those IT systems for the beef supply chain.

CHAPTER FOURTEEN

NOW AND THE FUTURE

As ABHA continued on firm footing, we began to have more time for ourselves and our ranches. Now our role in ABHA is "The Founders" and we appreciate the respect we are shown by the current members. Most have no knowledge of the shaky start of our association. Joe and I are always happy to talk about the history of the breed, when asked. We all must remember our history to keep from repeating mistakes—that's true in any part of life.

As we age, we have been fortunate to have our son, Dirck and his family come back to the ranches. So many ranching families don't have anyone who wants to work hard and take over. Many saw how difficult a life agriculture can be as they watched their parents. We are thankful for Dirck and Natalie, his wife. They have embraced the ranch lifestyle, they both work hard, and are raising the next generation of cowboys in the family, our two grandsons, Dayton and Reed. Dirck is a very well-respected member of the ABHA board and keeps us up to date on the status of the breed we founded. We are proud that Natalie was asked to join the Kansas Beef Council and is an active participant in their projects. They are doing good work in our community and with other ag organizations, such as Our Common Ground and BoysGrow of Kansas City.

Today our sale auctioneer is Andrew Sylvester, a contemporary of Dirck's. The next generation has rightfully taken over. Andrew was a young ring man when Stanley Stout, our first auctioneer, sold bulls for us in the 1990s for a $1400 average. Today, Andrew sells over $1 million worth of breeding cattle each year for us in our two annual sales (See Appendix D).

Joe and I now get to choose which chores we want to do at the ranches. Lucky us!

One of our favorite things to do is bull deliveries after our February Bull Sale each year. We love driving across America, this beautiful country of ours, to see our customers in their own places. Changes in landscape vary so much

J&N Ranch Inducted Into Hall of Fame

Joe and Norma Hoagland of J&N Ranch were inducted into the American Black Hereford Association's (ABHA) Hall of Fame on November 9, 2017, during the association's Annual Meeting. J&N Ranch is located in Leavenworth, Kansas.

Joe and Norma Hoagland registered the first Black Hereford, purchased the first ABHA membership, published the first Black Hereford magazine and held the first ABHA National Sale. Over 75 newspaper and magazine articles have been written nationally and internationally about the Hoagland's' pioneering work in the cattle industry.

The Hoagland's are also one of the largest independent citrus growers in Texas, producing Rio Red grapefruit as well as Navel and Valencia oranges near McAllen, Texas.

Joe is a graduate and former board member of Tulane University. He received a law degree from the University of Kansas in 1974. He served six terms in the Kansas State Legislature and was Majority Floor Leader in 1983 and 1984. He also served two terms as a member of the National Commission on Uniform State Laws.

26 The American Black Hereford | Spring

American Black Hereford, Spring 2018, announcing our induction into the ABHA Hall of Fame. Pictured are left to right: Joe, Norma, Natalie, Dayton, Dirck, Reed, and herdsman Brandon Furr.

The auction arena at J&N Ranch, Leavenworth, Kansas.

across America and each place is different. Each has its own beauty and challenges.

We have delivered bulls from Virginia to California, from Canada to southern Texas. Our Black Hereford cattle thrive from high in the Utah mountains to the bayous of lower Louisiana. The versatility, docility, and feed efficiency of Black Herefords are such strengths! On these delivery trips, we have been invited for lunches, dinners, and drinks by so many, but usually we are mindful of our passengers (the bulls!) and want to get them to their new homes as soon as we safely can.

Our best bull delivery trip and story, though, is to Hawaii. Well, of course, we didn't drive these two yearlings all the way to Hawaii! But Joe and I drove them to Los Angeles International Airport (LAX) where they would be loaded onto a 747-cargo plane. We drove them all the way, fed and watered each in their separate compartment, from Kansas through Oklahoma, Texas, New Mexico, Arizona, into California to our final drop-off at LAX Airport.

My favorite stop along that way was a historic motel, El Rancho Hotel and Motel, in Gallup, New Mexico. It was a drift back in time, staying where Humphrey Bogart and John Wayne had filmed their movies. The rooms were marked with the name of the film star who had previously stayed there. The lobby was full of movie photos, all autographed by stars we remembered. Moreover, our trailer with handsome bulls parked in the approved spot next to the street attracted lots of attention.

Then driving through Winslow, Arizona (while singing the appropriate Eagles song), we got to Los Angeles and our hotel. We drove up to the main entrance with trailer and bulls. Joe stayed with the rig while I went to the lobby, signed in, and got our keys. As I was coming back out the door, I heard people laughing and wondered. . . . Joe had told them we weren't staying because they didn't allow pets in the room! Traveling with bulls can be lots of fun. It can also be stressful if everything doesn't go according to plan. Those are other stories, not as much fun. But for this unique delivery trip, we had planned ahead, and all had gone well.

We rested until about midnight, then started for the airport, ready to make our required 2:00 a.m. time slot for delivery. It was amazing. There were stud horses, goats, all manner of livestock lining up for loading. We had fun visiting with other farm folks, right there in the middle of LAX, all sending our genetics off to Hawaii and beyond. Our two bulls had their own compartment, lined with hay and buckets of water. They loaded easily and we waved goodbye!

When the plane was underway, the young vet tech assigned to our bulls sent us cell phone photos of the two of them, calm and resting comfortably in their private cargo pen. When they arrived, we got another photo of the bulls with palm trees and the beach beyond. What a life they were going to lead!

Another of the chores Joe and I choose to do, as the older generation, is to work and stay at our remote ranch in the beautiful Flint Hills of Kansas.

Our Gunbarrel Ranch has always been a great source of pride and solace for us. It is named Gunbarrel because it encompasses Gunbarrel Hill, the highest

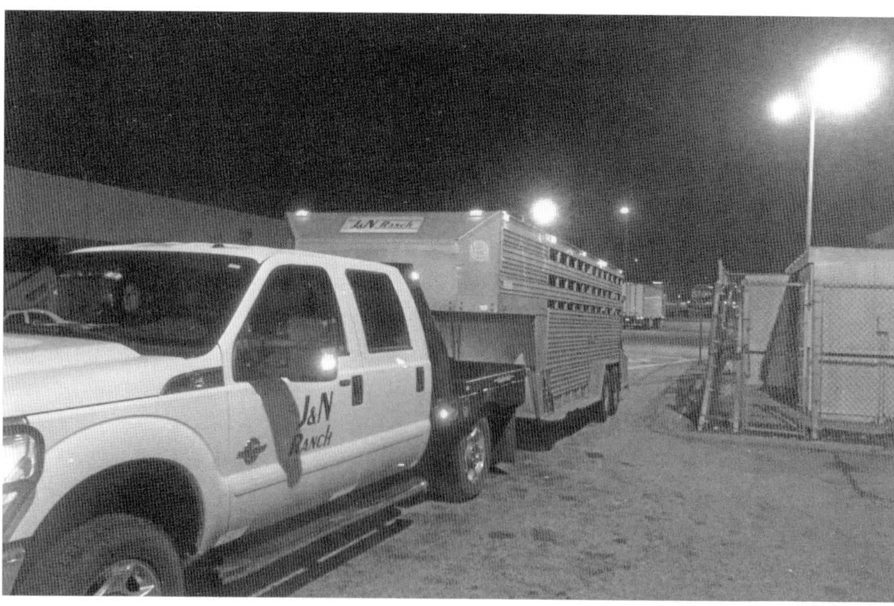

Unloading bulls bound for Hawaii at the Los Angeles Airport.

The first Black Hereford bulls arrive in Hawaii.

point in Wabaunsee County and second highest in the entire Flint Hills ecosystem. Located on the west side of the ranch, the road going over Gunbarrel Hill dates back to the Treaty of 1825. This was the first of several treaties between US Commissioners for Indian Affairs and the leaders of the Osage tribe. Later treaties were signed with the Kaw and Kansa tribes, also native to the Flint Hills. Gunbarrel Hill was crossed over by the trail the party followed from Fort Leavenworth across the plains to meet and sign the first treaty in Council Grove, Kansas.

With pre-Columbian views, this expanse of native grass breathes calm life into anyone quiet enough to experience it. We have joined several groups to preserve this precious part of Kansas, untouched by a plow. Only 4 percent of the original Tallgrass Prairie is left in North America and we are so fortunate to own and care for our portion of it. As Joe likes to say, it's only grass and barbed wire.

Although the view from the Missouri River bluffs at our Leavenworth ranch is stunning, I admit my favorite place is our small log cabin at our Flint Hills Ranch. The calm, the quiet, the never-ending glory of sunrises and sunsets out there are food for the soul. I listen to the birds and see what new native wildflowers are blooming each week during the growing season. The grass is like magic for our cowherd and the water from the springs runs clear and cold. Riding through the pastures, we see white-tailed deer, badgers, skunks, and sometimes scare up a burrowing owl who comes bursting out of his underground home to surprise and delight us.

Pairing with the Kansas Parks and Wildlife Department and the US Fish and Wildlife Service, we embarked on a prairie chicken habitat restoration project. Prairie chickens live and nest only where there are no trees, because predators for their chicks watch from trees, that is, eagles and hawks. Originally, of course, there were no trees on the vast Flint Hills, but with over-grazing and human and tree encroachment, trees have moved into the Flint Hills. We undertook a project to remove nearly every tree from our pastures, saving only trees around our small rustic log cabin for songbirds. We keep to a rotation of prescribed burning to keep the prairie healthy but give the birds time to nest and fledge their young.

Once we began to remove the trees from our Flint Hills pastures, I remembered a most satisfying conversation. Bob Widau, in his 80s, was one of the first real "pasture men" we had met at our Gunbarrel Ranch. He managed large acreages near us in Wabaunsee County. He found us one day, in our pastures, and complimented our tree clearing work, saying, "Your pastures look like I remember it looking out here when I was a boy." To me, that felt like vindication that we were doing the right thing to preserve the Flint Hills.

The results were satisfying in the first years. We hope our work there will continue to encourage these magnificent birds to keep their populations thriving. A wildfire in early 2025 set back the nesting population, but we remain

Joe and Norma at the shipping pens at our Gunbarrel Ranch, near Eskridge, Kansas.

hopeful and saw a good number of birds during summer and fall of 2025.

On a personal level, my energy had moved from the youth organization of 4-H, as our children grew and left home. I had loved my years as a project leader, community leader, and on the Board of Trustees of the Kansas 4-H Foundation. At the same time, I had begun to get involved in the alumni group for Watkins Scholarship Hall at Kansas University, where I had lived during my college years. I was raised in a family that "gave back" and always worked in organizations to benefit others. Now, I had more time to do just that, in our alumni group named Kitchen 8.

Watkins Scholarship Hall is an amazing story. The first of its kind in America, it provided housing for young women attending the University of Kansas in 1926. This was a time when there were few women attending college and even fewer proper housing options for them. I was lucky enough to live in Watkins and as a part of our alumni group, we are always looking for ways to make others more aware of this gift. One hundred years later, this lovely building is still providing affordable, sustaining housing for young women attending KU. The hall was built by Mrs. Elizabeth Miller Watkins, a fascinating woman who designed this unique, scholarship dormitory.

I became aware of a series of books written about landmark buildings on the campus at KU. It was a bit cheeky, but I asked for a meeting with the sponsoring group of this book series. I made a presentation about Watkins Hall and its' identical twin, Miller Hall, and why they should be included in the book series—and they said yes. Now I was faced with the task of writing a book, at the age of sixty-five, when I'd never written a book before. As always, Joe encouraged me to take on this new challenge. Why couldn't I do it, he asked of me? With the help of the editors of the three previous books, Carol Shankel and Barbara Watkins, Watkins and Miller Halls was published in 2016.

Doing the research for this architectural history book had one problem, though. I got so hooked on the life of Mrs. Elizabeth Watkins that I kept on researching. Reading her letters, housed in the Spencer Research Library at KU, I was captured by this strong woman. She was a woman in business in the 1890s when women weren't in business! At one point she and her husband, J. B. Watkins, were among the largest landowners in America. Owning vast acres of Kansas farmland, they built their own railroad to Louisiana to ship grain from Kansas fields to a Gulf port. Big Bend National Park in Texas was part of their land holdings. They developed the rice industry in Louisiana, on some of the millions of acres they owned there, designing irrigated fields near the town of St. Charles.

How had Elizabeth followed this path? How had she been so successful? How did she manage to defy social norms to live in New York City as a single woman, running companies for J. B. before they were married? Eventually, I paired my research with that of Mary Dresser Burchill, a local Lawrence historian. Mary had also been doing research on Elizabeth for years. We decided to write Elizabeth's biography and University Press of Kansas published it. As we combined our work, research, and skills, we laughed that Mary was the "play by play" and I was the "color commentary."

The Life and Legacy of Elizabeth Miller Watkins was published in 2023.

My work with my alumni group continues as we plan a celebration for Watkins Scholarship Hall's one-hundredth anniversary in 2026. The building is still there, still housing young women, one hundred years later. At KU, following Elizabeth's idea, there are now twelve scholarship houses. It is a fitting tribute to the woman who designed, furnished, and built the first two.

Joe and I both love the satisfying work we have done throughout our lives. We also realize we are reaching an age when clearly, we must pass on our enthusiasm and projects to the next generation.

We learned from our European friend, Thierry de l'Escaille, that three things are needed to effectively pass on a family business to the next generation. First, it must be profitable. Second, the next generation must want to work at it, not play at it. And third, the older generation must get out of the way. We like to think we have done all three with our family businesses.

The entire Hoagland family. (Photo by morgan jane photography, Leavenworth, Kansas)

After six generations, the Hoaglands have become adept at generational transfers of family businesses and Joe was laying the groundwork throughout our life together. With three children and three different family businesses—airline equipment sales, cattle ranching, and growing citrus—it was helpful to give a separate business to each child.

Racket Merchandise Company, the airline equipment company and historic source of so much of our family wealth, is in the capable hands of our eldest daughter, Anne, her husband Brandon, and our granddaughter Lillie. Anne had worked at Racket since her high school days. Now she has also begun a separate business, Legacy, as a coach, adviser, and lecturer on family businesses and intergenerational transfer. We are proud of her work, helping other families with their lifetime journey. Along with lectures and presentations, she also publishes a magazine "Legacy."

Today, Lone Star Citrus, in the Rio Grande Valley of Texas, is focused on growing oranges and grapefruit. This is a change from the investment purpose that Virgil had when he originally bought and sold groves for housing developments. As we turn the management of Lone Star over to our younger daughter, Emma, we three believe in the future of agriculture and growing food in the USA. In 2026, Lone Star will become the second largest organic citrus grower in Texas, producing Rio Red grapefruit, Navel, and Valencia oranges for Whole Foods and other organic grocers.

Lone Star Citrus owns our smaller and newest ranch in Fair Play, Colorado. It is so different from the other ranches, at ten thousand feet and next to the Pike National Forest. The forests are full of herds of elk, moose, and black bears. It's not a place we go for a walk after dark! It has magnificent views, clear air for hiking, and a way of renewing our love of nature. We are so proud of our younger daughter, Emma, who maintains a career as a pediatric nurse in Denver but also manages this ranch for us, along with overseeing the Lone Star citrus production in Texas.

We spend time at our Colorado ranch at least twice a year. Of all our homes, the ranch in Colorado has a special draw for Joe. As he said to me, "Although it's the least profitable of the ranches, Colorado has always been a summer destination for the old families of Kansas City. My great-aunts and uncles and great grandparents all vacationed in Colorado or owned summer homes in the mountains to escape the heat and humidity of Kansas City summers, before air conditioning. I have old photo albums showing the family camping along the roadsides on a two-lane road en route to a mountaintop getaway. I think of the Colorado ranch differently because of that family history connection, even when other properties have been owned by the family longer."

LEGACY

SPRING 2025| ISSUE NUMBER 3 FAMILY BUSINESS MAGAZINE

Healthy Boundaries for Healthy Relationships in Family Business

Boundaries feeling a little messy or leaky? You're not alone.
Healthy boundaries allow us to care for ourselves and others. This can look different in each relationship because people have unique needs and dynamics - *especially* in family businesses.

How to Start the Conversation in Multi-Generational Family Businesses

5 Questions with 3rd Gen. Doll Distributing Beverage Co.

The Energetics of Family Structures & Roles in Family Business

Cover of the family business magazine, Legacy, published by Anne Hoagland Bauer.

Photo of our organic citrus orchard in the Rio Grande Valley of Texas.

This story of Joe's and my life's work won't end, as the work we've done as a team will continue. J&N Ranch, the Birthplace of Black Herefords, with our son Dirck and his family at the helm, is in caring and capable hands.

We are so very fortunate to have children who have seen our work, have absorbed their family history of work, and are making solid, productive, and happy lives for themselves.

At this later stage of life, we realize how much all our experiences in politics, law, business, advertising, travel, and even beauty pageants prepared us for this adventurous life we've led. Every new day has a purpose for us. Getting a good basic education made us aware we always had more to learn, and we could keep learning all through our lives. This enabled us to start the Black Hereford cattle breed, which is likely our most durable and far-reaching achievement. (Well, unless Joe can actually save the Amazon Rainforest with CattleTracs!)

Two people, able to work together with respect and love, can accomplish so much more than each could separately. Another of my favorite authors, Jane Austen, wrote, "Happiness in marriage is entirely a matter of chance."

Well, "chance" smiled on us as we found each other and each had a family legacy of hard work to fashion our lives around. It's been a joy to share this life with you.

Joe and Norma at our Lone Star Ranch, near Fair Play, Colorado.

APPENDIX A

BLUE JACKET HEREFORDS
JOHN AND DAVID GAGE
1282 E. 1900 RD.
EUDORA, KANSAS 66025

February 5, 1994

Mr. H. H. Dickenson,
Executive Vice President
American Hereford Association
P.O. Box 014059
Kansas City, MO 64101

Re: American Black Hereford Association

Dear Hop:

I enclose herewith a copy of the Articles of the above, which have been recorded in Kansas. The name has been reserved in Missouri. No rules have been drawn up.

We have no intention of promoting such an association. Its purpose is primarily defensive and for the time being at least it will be private and not open to all breeders. It will be used to develop a few black whiteface cattle of what we consider the best available breeding to compete with the better registered Angus bulls that at this time are winning the competitive battle for bull sales in our area.

I do not have the confidence you and some others have in changes that will result from the black cattle being offered by continental breeds. It could have the effect of pushing a lot more commercial breeders toward straight black Angus operations unless they can see a clear alternative, as has been indicated by conversations I have had with several successful commercial producers.

I enclose copies of pages from the Select Sires 1994 book with pedigrees and details on the two Angus bulls we have semen of on hand. There are too few, if any, Hereford bulls available that present this kind of data. The same is applicable to a comparison we see between many Hereford and Angus bulls being offered at auction. We have seen a trend in the right direction by some Hereford breeders recently, but they (including us) still have a long way to go. We intend that this new herdbook be used for low birthweight, higher milk, better-uddered, more pigmented and more fertile cattle with better carcass value and will write rules to that effect. We plan to put the calves in your TPR and try to develop a formula for some realistic EPDs that will compensate for the gap from the Angus cross.

I was shocked last summer when I attended a Polled Hereford seminar and heard two respected college experts (beef genetics and meats) express an opinion that Herefords would not play a significant role in the make-up of the beef cow herds of the future. And that they would be predominantly Angus or Angus crossed with breeds like Shorthorn or Charolais. And that the role of Herefords would largely be limited to a terminal-cross type use. They did concede that attributes of Herefords would give them value in that sort of role, making reference to the Colorado State - Monfort study.

Frankly I have no desire to be relegated to terminal-sire breeder status and I had hoped to see some light and get some inspiration from the report of the policy direction committee in Kansas City, but did not find it. It was sort of a "deja vu" thing for me, recalling to mind the American Milking Shorthorn Society when I was elected president over 30 years ago. Its registrations were suffering a drastic decline while some breeders were keeping shows at high levels, socializing, and bragging about their "breed that fills every need", without really analyzing what their competition was doing and what they would have to do to meet the challenge.

Yours very truly,

John

John C. Gage

APPENDIX B

John C. Gage Law Firm

Attorneys & Counselors at Law

John C. Gage
John B. Gage II (of Counsel)

LAWRENCE KANSAS OFFICE
1009 NEW HAMPSHIRE
SUITE A
LAWRENCE, KANSAS 66044
TELEPHONE (785) 865-0766
FAX (785) 843-7785

COFFEYVILLE KANSAS OFFICE
121 WEST 8TH STREET
COFFEYVILLE, KANSAS 67337
TELEPHONE (316) 251-5950
FAX (316) 251-5955

November 6, 1997

Mr. Joe Hoagland
J & N Ranch
25332 Wolcott Road
Leavenworth, Kansas 66048

Re: American Black Hereford Association

Dear Joe:

We are delivereing herewith all our records on the above corporation. These include minutes of a meeting November 5, 1995, in which the registered office was changed to your address with you as registered agent. The certificate that needs to be filed with the Secretary of State is also included, executed by David and myself. I think the filing fee is $20.00, the same as the annual fee with the next annual report due in 1998. You are supposed to record the certificate the Secretary of State sends back in your Register of deeds office.

As you will note we did not adopt bylaws because they have to be tied in to the rules of membership and registry. I include in the enclosed some reference material you may want to look at in developing such rules. It would probably be advisable to get rules from other associations like the Brangus, Gelbvieh etc to see how they have handled setting up a registry for crossbreds. My tentative idea about it was to record any Angus-Hereford crosses with registered ancestry, then have them move to full registry when they were perhaps at least 5/8ths Hereford, with a special registry for those in full registry that are proven by DNA testing or otherwise (a lot of progeny) to be homozygous black. Some performance requirements should also be involved, maybe reporting weights on all calves.

The annual report includes $100 cash assets but we have used that up for filing fee and annual report fees, so the treasury is bare.

Call me if any questions. Will be real interested in what you do with it. Am excited about the black 51M bull calf out of the good Traveler 5204 daughter.

Yours very truly,

John

APPENDIX C

BLACK HEREFORD vs ANGUS

HEAD-TO-HEAD COMPARISON

We purchased these two sets of yearling steers in the Spring of 2022. They were turned out on grass at our Flint Hills ranch. We shipped them off grass in the Fall of 2022 directly to the feedlot. They were fed in adjacent pens exactly the same rations and sold on the U.S. Premium Beef grid in February 2023.

The Black Hereford steers were purchased from a customer who has been using our bulls for the last 11 years. The Angus steers were off a reputation Angus herd and were also bred alike.

AS STOCKERS	BLACK HEREFORD	ANGUS
# of yearling steers	66	74
Delivered purchase price	$1278/hd	$1308/hd

	AS FEEDERS	
Weight in	835 lbs.	809 lbs.
Weight out	1495 lbs.	1398 lbs.
Feed cost	$47,307	$47,972
Deads	1	3
# of cattle out	65	71
Gain/hd	647 lbs.	555 lbs.
Cost of gain/lb.	$1.13	$1.22

SOLD TO U.S. PREMIUM BEEF ON GRID		
Prime	13.8%	2.9%
CAB (Upper 2/3 of Choice)	32.6%	25.8%
Choice	96.8%	97.5%
Select	3.2%	2.5%
Total Sales	$158,909	$155,030
Sale Price/hd	$2,445	$2,184

OVERALL	BLACK HEREFORD	ANGUS
Cost/hd	$1278	$1308
Feed cost/hd	$728	$675
Sell price/hd	$2445	$2184
PROFIT	**$439/hd**	**$201/hd**

APPENDIX D

SALE REPORT

J&N Ranch's 42nd Production Sale

February 14, 2026

Leavenworth, Kansas

Auctioneer: Andrew Sylvester

Averages:

115 Black Hereford Bulls $9,295

Total $1,069,000

J&N Ranch held their annual bull sale on a rainy Valentines Day at their ranch a 30-minute drive north of Kansas City. Buyers from 15 States purchased from a large selection of coming two-year-old Black Hereford bulls ranging from $6,000 to $25,000. The sale also featured a group of select yearling bulls selling from $7,000 to $17,000.

Among Black Hereford breeders, the high selling bull was purchased by David Unterbruner of South Dakota edging out second place bidder Brock Johansen of Utah for JN Balder J009M for $25,000. Schroeder's Black Herefords of Missouri paid $15,000 for JN Balder 808N and Logan Hill of Nebraska paid $15,000 for JN Balder H055N both popular yearling bulls. Earlene Boone of Arkansas purchased JN Balder 708 M, a stylish two-year-old bull for $14,000.

Our next sale is October 3rd featuring a few select bulls, 3 in 1 pairs with heifer calves at the side and bred heifers.